The Wildlife Trusts Handbook of
GARDEN WILDLIFE

Nicholas Hammond
Foreword by
Chris Packham

B L O O M S B U R Y

LONDON · NEW DELHI · NEW YORK · SYDNEY

First published in 2002 by New Holland Publishers (UK) Ltd
This edition published in 2014 by Bloomsbury Publishing Plc

Bloomsbury Publishing Plc, 50 Bedford Square, London, WC1B 3DP

www.bloomsbury.com

Bloomsbury is a trademark of Bloomsbury Publishing Plc
Bloomsbury Publishing, London, New Delhi, New York and Sydney

A catalogue record for this book is available from the British Library.

ISBN (print) 978-1-4729-1586-3

Printed and bound in China by C&C Offset Printing Co., Ltd.

10 9 8 7 6 5 4 3 2 1

Acknowledgements
Thanks must go to all the authors to whose books I have referred for
information. My particular thanks go to Ian Dawson, the Chief Librarian
at the RSPB, whose knowledge of natural history information sources is
breath-taking, and to my Wildlife Trust colleague, Brian Eversham, who
was able to answer some of my queries on invertebrates. Working with
New Holland has been enjoyable, for me at least, and I thank Jo
Hemmings for having the wisdom to commission this book, Mike
Unwin, the project editor, for his patience, enthusiasm and creative
contribution, and to Sarah Whittley and her talented team of
illustrators, for their invaluable contribution.

Contents

The Wildlife Trusts

The Wildlife Trusts are the UK's largest people-powered organisation caring for all nature - rivers, bogs, meadows, forests, seas and much more. We are 47 Wildlife Trusts covering the whole of the UK with a shared mission to restore nature everywhere we can and to inspire people to value and take action for nature for future generations.

Together we care for thousands of wild places that are great for both people and wildlife. These include more than 760 woodlands, 500 grasslands and even 11 gardens. On average you're never more than 17 miles away from your nearest Wildlife Trust nature reserve, and most people have one within 3 miles of their home. To find your nearest reserve, visit **wildlifetrusts.org/reserves**, or download our free Nature Finder iPhone app from the iTunes store. You can also find out about the thousands of events and activities taking place across the UK – from bug hunts and wildplay clubs to guided walks and identification courses – on the app or at **wildlifetrusts.org/whats-on**

We work to connect children with nature through our inspiring education programmes and protect wild places where children can spend long days of discovery. We want children to go home with leaves in their hair, mud on their hands and a little bit of nature in their heart. Find out more about our junior membership branch Wildlife Watch and the activities, family events and kid's clubs you can get involved with at **wildlifewatch.org.uk**

Our goal is nature's recovery – on land and at sea. To achieve this we rely on the vital support of our 800,000 members, 40,000 volunteers, donors, corporate supporters and funders. To find the Wildlife Trust that means most to you and lend your support, visit **wildlifetrusts.org/your-local-trust**

The Wildlife Trusts
The Kiln, Mather Road, Newark, Nottinghamshire NG24 1WT
t: 01636 677711
e: info@wildlifetrusts.org

Registered Charity No 207238

wildlifetrusts.org

Find us on
Twitter - @wildlifetrusts
Facebook – facebook.com/wildlifetrusts

Foreword

There is no clearer indication as to the value of our gardens as a resource for wildlife than the diversity of exciting creatures identified in this book. Yes, this myriad of creepers, crawlers, slitherers, scuttlers and flappers could be no more than a stroll out of your backdoor! So forsake your television, your newspaper or your afternoon nap, arm yourselves with children if you have them, pull on a pullover and set off on a real-life safari. Okay, some of the species are pretty small, not so fierce or even that beautiful but they are on your doorstep and playing a role in the community which you share. You ought to get to know them.

Now, we all know Foxes, Blue tits and Ladybirds, or at least we think we do. But simply recognising or identifying is not really 'knowing' and more time spent studying even familiar species is inevitably rewarding and enlightening. So for me the strength of this book is its scope – it is a great guide to the less familiar, to the overlooked and under-rated animals which rarely secure a second glance. But here is your opportunity – a clear, concise and precise method of locating and identifying a whole new fauna.

The illustrations are great, the instructions friendly and the information pertinent – this is a book for both young and old and for all biological abilities, even the most accomplished naturalists will find something new here. And when you've been excited by an ant, an aphid or an accentor you will want to leap over the garden fence and into the countryside in the search of less urban and more urbane species. No better destination then than your local Wildlife Trust Reserve. Join now, avoid the queues and play a valuable role in the wider aspects of conservation! Britain has a great tradition of producing brilliant amateur naturalists and this book will be a valuable asset to the next generation of this species.

Chris Packham

Introduction

Gardens in northern Europe are wonderful places to start watching wildlife. Gardens vary from tiny plots in towns to country gardens of several hectares. The larger the garden the greater the array of wildlife is likely to be. This book is intended for the person who wants to try to identify the animals that appear in the garden. These may be so small that, as is the case with many insects, a lens is needed to see them more closely or they may be as large as a badger or a roe deer. To see birds or larger flying insects, such as dragonflies or butterflies, you may find it easier to use binoculars. There are several ways you can encourage wildlife into your garden, I have mentioned a few ways on page 11. It can be very frustrating trying to identify animals, especially ones that fly. Even mammals can be troublesome, as most are nocturnal and very secretive. The best way of observing shy mammals is to make or buy a hide. Once you know you have visitors in your garden, sit very still and wait until they think you have gone, you will be amazed at how close they will come.

This book cannot be comprehensive since there are so many species that might turn up in gardens. Animals that can fly are the obvious accidental arrivals. Birds blown across the Atlantic or having hitched a lift on a ship may appear in gardens in the west of Europe. I remember seeing a laughing gull, a North American bird, sitting disconsolately on a fencepost behind a butcher's in Shetland, and vagrant birds are not uncommon visitors to gardens in the Scilly Isles. The proximity of countryside will add to the species found in gardens. My own garden list includes an osprey which caught a fish in the river that borders our garden. This bird was migrating southwards in September on its way to its winter quarters in West Africa and the sighting lasted less than a minute.

The animals that are seen in gardens will vary from season to season. It is well-known that some birds migrate southwards in winter and return to breed in spring. However, several species move into gardens in search of food in winter and then disappear in spring, when they return to woodlands and other habitats to

Although generally associated with more rural habitats, badgers may be encountered in suburban gardens, where they supplement their staple diet of earthworms with choice scraps from the kitchen.

breed. Other animals also disappear from view during some seasons. This does not necessarily mean that they have left the garden: they may be hibernating or continuing their lives away from our view.

The animals that have been included are amongst those most likely to be seen in gardens and this book will also help you to identify some animals you come across in the countryside. At the end of the book is some advice on other sources of information about the wonderful wildlife it is still possible to see in Europe today.

Secretive and often overlooked as they forage on the ground for invertebrates, dunnocks are best observed in gardens. They draw attention to themselves in spring with a melodious song uttered from a prominent perch.

Habitats and distribution

Animals have over millions of years adapted to make the most of their particular environment. In the case of some species of invertebrates the habitat to which they have adapted is small and specialised, being confined to a certain food plant and having specialised requirements for breeding. Generally, such specialisation leads to rarity, and, therefore, it is usually the generalist species that are found in gardens, because they can survive in a variety of habitats.

Although gardens are man-made, they will have obvious traces of the habitats which they have replaced. Where houses have been built in what was once woodland, the gardens sometimes contain mature oak trees, which attract species of birds and insects not found in other gardens. Where houses have been built on farmland there may be the remnants of hedges, with their characteristic woodland edge fauna. Because many species, especially those that fly, are often very mobile, all sorts of creatures may turn up in our gardens. It is worth knowing what the habitats nearby might contain. For this reason we have included details of the habitats in which the animals are usually found.

The garden is usually like the understorey on the edge of a wood or a highly modified piece of meadowland. Gardens do vary according to the surrounding countryside. Soil is an important factor, particularly where invertebrates are concerned, because they are often linked to a particular species or group of species of plants, which in turn may be linked to a particular soil type. The habitats that surround the town in which the garden lies or the countryside in which the town lies will affect the species that are seen in the garden. A garden on the edge of heathland is far more likely to attract heathland species than one set in the middle of arable farmland. Although it is possible, if not probable, that any species of bird might turn up in a garden, because birds can fly. Most invertebrates are seldom seen far from the habitat in which they specialise.

Very few species that come into gardens in Europe visit gardens in the United States. This is because the movement of continents has isolated animals and they have evolved differently in different parts of the world. Thus, just as marsupials, such as kangaroos, are found only in Australia, hummingbirds are only found in the Americas. There are six of these biogeographical realms: the Nearctic (North America), the Neotropical (Central and South America), the Ethiopian (all but the north-western corner of Africa), the Oriental (the Indian sub-continent and south-east Asia to Borneo, the Australian (from Sulawesi to New Zealand and including New Guinea), and the Palearctic (Europe, Asia and north-west Africa).

Northern Europe is a tiny part of the Palearctic realm, but even within there are variations in the distribution of species. These are caused by the availability of food and by the geography of the area. In the far north towards the Arctic Circle, snow and ice will cover the land for much of the year and therefore there is only a short period during which a limited number of species can survive. In autumn as the weather becomes harder and food scarcer, mobile animals, such as birds move south. Geography and climate change have played their part in the distribution of some animals. The islands off the north-west coast of Europe have fewer species than the mainland, despite once having been attached to the mainland. When much of Europe was covered in snow and ice during the last Ice Age, Britain and Ireland were still part of the Continent, and as the Ice Age declined, about 10,000 years ago, it would have been possible to walk from southern Spain to Shetland. As the climate became warmer, many species of animals moved north, taking advantage of the woodlands that were taking the place of the tundra and ice-covered terrain, but the warmer weather melted the ice and raised the sea level, cutting off Britain from the mainland and Ireland from Britain. This made it impossible for any more species of non-flying invertebrates, amphibians, reptiles and mammals (except for bats) to cross the English Channel. Therefore, only those species that had already made it or which were introduced by man, either deliberately or accidentally, live here.

A mature oak tree in your garden offers a treasure house for wildlife. Oaks are associated with over 500 species of invertebrate and an unrivalled diversity of fungi, as well as providing food and nesting habitat for an abundance of birdlife – including woodpeckers, owls and jays.

Looking at animals

Both size and colour of animals can be difficult to estimate. For example compared with a buzzard, a sparrowhawk is rather small, but when compared with a sparrow it is quite large. Similarly, butterflies can look larger in bright light, although males and females and individuals may vary in size. Colour, too, can vary. Birds' plumage can alter colour as the feathers become worn before its moult. It is the effect of light that is the major cause of confusion. Local colour will affect the appearance of the bird's colour: the colour of the breast of a robin next to a red watering can will look different from a robin on the soil in the shade of a laurel bush. The quality of light will alter the colour of a bird. Thus, if the natural light is behind a usually colourful bird such as a goldfinch, it can appear to be quite dull, while a house sparrow bathed in evening sunshine will appear so colourful it might have come from the Tropics. Looking through this book you will begin to build up an idea of what each species looks like and this will help when you see it in your garden, but it is helpful to concentrate on the animal before and impress its image on your mind.

Swallows are seasonal garden visitors. They arrive in Europe in spring, once the warmer weather produces a plentiful crop of their insect food, and fly south again in the early autumn. Swallows may breed in rural gardens, but in towns are more commonly seen passing through.

Food and behaviour

The fundamental aim of the life of an animal is survival. This may be the survival of the individual or of its genes, the survival of the species. Food, shelter and reproduction are the keys to this. The requirement for food is obvious and, since this is also the area where gardeners and wildlife are most likely to find themselves at odds, these are important factors in identifying the species. Some of the animals that come into gardens can be pests, because they eat the plants grown by gardeners. Others, on the other hand, could be encouraged because they feed on the pest species and provide at least some control over their numbers.

The slow worm, a legless lizard, is a blessing to the gardener since it feeds almost entirely on slugs. Slow worms can be encouraged in the garden by providing appropriate refuges in the form of large stones or old logs.

The behaviour of animals may also be a clue to their identity. Animals appear at different times of day and in different seasons. An obvious example is the swallow, which appears in March or April and disappears in September or October, because it is a migrant. The swallow breeds in Europe and when the insects it eats become scarce, flies south for the winter. Other birds may change their feeding habits to cope with the changes of food from season to season. Blackbirds, for example, concentrate on earthworms, insects and other invertebrates when they are plentiful in the summer, but will turn to berries and fruit in autumn and winter. It is in winter when food is scarce that birds will be attracted to gardens. Great tits and blue tits descend on gardens in winter in large numbers, making the rounds of garden bird feeders. In spring, most move away from gardens to breed in woodlands, where there are plenty of insects to provide food for their young. Some of the other species of birds, such as fieldfares and redwings, that visit gardens in winter are migrants from their breeding grounds in Scandinavia and eastern Europe.

Butterflies are seen during the day and most species of moth fly at night. They also occur seasonally. The peacock is a common garden butterfly, especially if there are nettles on which the caterpillars might feed. The first peacocks are seen as the weather warms in March and they emerge from hibernation to breed. After May they seem to disappear, because the adults have bred and the new generation is going from eggs to larvae to pupae to emerge as butterflies from July to September. On sunny autumn days they will be seen feeding on the nectar of the last michaelmas daisies or juices of rotting windfall fruit, before hibernating throughout the winter. In the depths of winter you might come across one in a corner of the garden shed or even in the house. The life cycle of all insects includes a number of stages, which take an insect throughout the year from egg to the adult insect. These stages vary between species in terms of the timing, but each cycle is timed to provide the maximum opportunity for the larvae to feed.

Reproduction is, of course, a vital part of the lives of animals. Birds sing in the spring to proclaim that they hold a territory and to attract a mate. The flight of male brimstone butterflies on warm spring mornings, has the same purpose. You will notice other butterflies and dragonflies patrolling your garden in search of mate. Having paired, birds find places in which to build nests and rear young, which means that there has to be sufficient food for their young. Insects have to find the food plants on which their larvae can feed or, if they are predatory, enough of their prey.

Brimstones are one of the first butterflies of the year to appear in gardens. Visible as early as March, the conspicuous bouncing flight of the male serves to proclaim his territory and attract a mate.

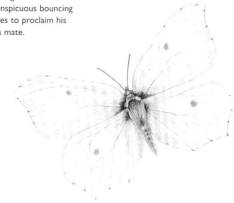

Making a garden fit for wildlife

A pristine garden, in which pesticides and herbicides are used extensively and where lawns are cut to provide a neat sward like a billiard table, is not likely to attract much wildlife. That does not mean that your garden has to be a wilderness in order to attract wildlife. Indeed, if a garden is left to run riot, it may eventually become attractive to a smaller variety of wildlife than a well-managed garden, in which consideration is given to wildlife.

If you garden in a non-harmful, "organic" way, avoiding using herbicides or pesticides, you will attract wildlife, even though you have a conventional, short-cut lawn and formal herbaceous borders. Discourage cats, which are the greatest killers of garden wildlife. During the breeding season, try to minimise disturbance to hedges, where birds nest, and to compost heaps, which provide refuge to a host of wildlife from hedgehogs and grass snakes to earthworms and centipedes.

To attract any wildlife a garden must have food and shelter. Obviously the larger the garden the easier it is to provide these, but it is possible to attract birds to the smallest garden by putting out food in the form of kitchen scraps and seeds bought from a garden centre, petshop or a specialist supplier. Insects, particularly bees, hoverflies and butterflies, can be attracted by planting bushes and flowering plants from which they can extract nectar and pollen. Birds will be attracted by berry-bearing shrubs.

Shelter in the form of nestboxes will attract birds, but they will only nest if there is sufficient food available and this means small insects, because even the seed-eating birds feed their young with insects. Many insects and other invertebrates require shelter when they are not active, so that under leaf litter there will be a large number, some of which will emerge at night to go in search of food or to find a mate or mates. Compost heaps, as well as being a positive way of reusing waste, are also an important place in which invertebrates will live and breed. A compost heap may also provide hibernation places for hedgehogs, grass snakes and toads. Shelter for insects can be provided in various ways. Hollow sticks held into a frame with chicken wire can provide holes in which solitary bees and wasps can nest. Log-piles will be used as shelter by many species. You do not have to have a wood-burning stove or an open fire to make a log pile. When you have to lop a branch from a tree, cut it into logs and pile them up. It will make an attractive "garden feature" and will save you a trip to the dump and leave some space at the landfill site.

Every wildlife garden should have a source of water in which birds can bathe and drink. The best way of providing this is to dig a pond. Even the smallest garden pond has value for wildlife. As well as being drinking places for birds and hedgehogs, ponds are breeding grounds for frogs, toads, newts and insects, which spend their larval stage in water. Do not succumb to the temptation to add goldfish to the pond, because they will eat other forms of life in the pond. It is remarkable how rapidly wildlife will colonise a garden pond: recording which species arrives and when, will add to your enjoyment of the wildlife in your garden.

A simple log pile left to decay undisturbed in a corner of your garden quickly becomes a mini eco-system of its own, providing food and shelter for a wealth of animal and plant life.

Naming and classification

Wildlife in the garden varies from primitive single-celled organisms to highly developed mammals, although none in northern Europe are visited by any of the most highly evolved mammals – the primates, except, of course, for man. This book describes some of the animals most likely to be encountered in our gardens and to be recognised without the aid of a microscope (although a hand lens might be useful with some species). It starts with the most primitive and progresses through the animal kingdom to the most highly developed.

To classify animals biologists divide them into a number of groups. Animals are divided into two main groups – invertebrates and vertebrates. Vertebrates have backbones and skeletons, this group consists of fish (not included in this book), amphibians, reptiles, birds and mammals. Invertebrates, which may, like earthworms, have no hard parts or like insects have a hard exoskeleton covering the soft parts of their bodies, cover every other class of animal. The next major category is the phylum, a very broad group and beneath that the class, which groups together all animals sharing similar characteristics: mammals form a class of their own. Within each class there is a major division into orders: rodents are an example of a class of animals. The next important category is the family which includes all species that share certain characteristics: mice are a family within the order of rodents. Within each family the next category is the genus, grouping together similar species. The species is the group that contains all those individuals that have similar characteristics and which can produce fertile young. Although species within the same genus may interbreed, they do not usually produce fertile young; for example, the offspring of a horse and donkey, which share the same family, is the infertile mule.

All animals are given scientific names, which are based on a combination of Latin and Greek. The name of a species has two words. The first is the name of the genus and the second is the species. For example, the scientific name of the blackbird is *Turdus merula* and the closely related song thrush is *Turdus philomelos*. We have included the scientific names of all the species in this book in order to show where species are related and to make their identification easier, because the English names may vary in different parts of the country.

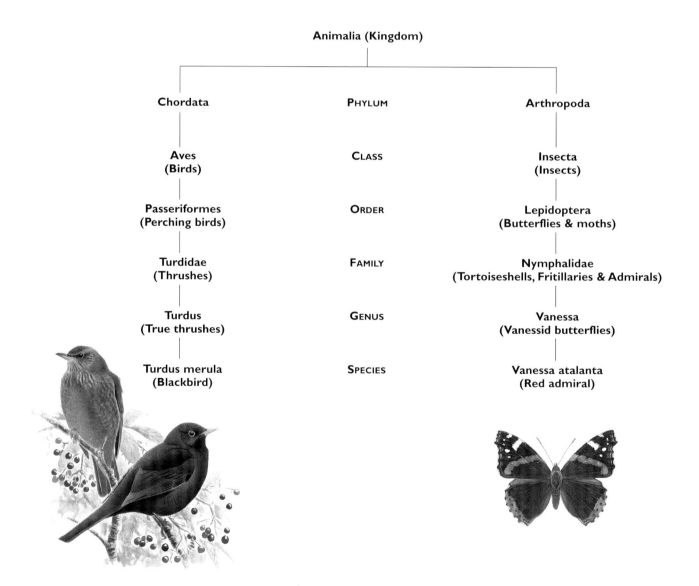

Animalia (Kingdom)

Chordata	PHYLUM	Arthropoda
Aves (Birds)	CLASS	**Insecta** (Insects)
Passeriformes (Perching birds)	ORDER	**Lepidoptera** (Butterflies & moths)
Turdidae (Thrushes)	FAMILY	**Nymphalidae** (Tortoiseshells, Fritillaries & Admirals)
Turdus (True thrushes)	GENUS	**Vanessa** (Vanessid butterflies)
Turdus merula (Blackbird)	SPECIES	**Vanessa atalanta** (Red admiral)

How to use this book

The key to identifying wildlife is looking and seeing. The better the image you have of the animal, the better your chances of identifying it successfully. In some ways it is easier to identify small, slow-moving invertebrates, because you can temporarily capture them and look at them at your leisure, possibly using a lens. However, larger, fast-moving animals such as dragonflies, butterflies and birds do not give you so much chance. Binoculars will help. For garden bird- and insect-watching I prefer to use those which can be focussed on an animal a few metres away. This gives me a chance to look at larger insects from a distance so as not to disturb them, and to examine in detail the plumage of birds at the bird-table. If a bird, butterfly or dragonfly that it is strange to you arrives in your garden, do not immediately rush for this book, because while you are flicking through the pages, you are not looking at what you are trying to identify. Stare at the bird, noting its shape, its colouring, the way it moves and, if it's feeding, how it pecks at the food. Having impressed these details on your mind, make some quick notes. Then, when you have amassed the information, turn to this book.

As has been explained the wildlife here is arranged with the most primitive species at the beginning and the most highly evolved at the end. For the invertebrates there is a general key and then once you have identified the probable class of invertebrate, for example insects, you will find there is another key at the beginning of the section on insects. There is another key for wasps and their allies that should take you to the genus and in some cases the species.

On the previous page is an explanation of how animals are classified. Each species has a scientific name, which is often in Latin or Greek or a combination of the two. Because English names may vary considerably we have included the scientific names. In the case of some of the invertebrates there is no English and, therefore, we have had to use scientific names.

There is an illustration of each animal described. The appearance of many species varies with the season and variations are described under the heading "Size and description". With smaller animals that can be measured in the hand, the size can be a useful indicator, but with larger animals, such as birds, the size can be very difficult to judge in the field and the real value of the measurement is to give you the opportunity to compare with other similar species.

The descriptions include an entry for "Habitat", which gives brief details of the type of habitats in which the species is found and its distribution within northern Europe. This can be an important feature, because the habitats in the areas surrounding your garden may give a clue as to the species you have found. There is also an entry for "Food/habits" and this covers the season at which the animal is most likely to be seen, its feeding habits and any distinguishing behaviour.

For birds there are additional entries for "Voice", an important distinguishing feature, and details of breeding.

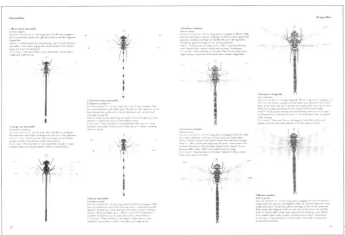

Colour illustrations clearly reveal key identification features. Each species is depicted in proportion to the others on that page. The accompanying text gives further tips for identification, including size, and provides a brief account of habitat and behaviour.

Invertebrates

The most primitive forms of invertebrates are the Protozoa, single-celled animals, which are microscopic. In this book the most primitive are the free-living flatworm, *Dugesia tigrina* which is found in water, the tapeworm, *Toxicara canis* and the nematode, the potato eelworm (page 17). Unlike the earthworms (page 16), which belong to the order Annelida, flatworms, tapeworms and nematodes have bodies that are not segmented.

Slugs and snails (pages 18–20) belong to the phylum, Mollusca and the class, Gastropoda, which means "stomach-foot" in Ancient Greek, a reference to their crawling on their bellies. They are soft-bodied and snails have thin, coiled shells, which grows as the snail's body grows.

The large phylum of invertebrates known as Arthropoda includes several classes whose members are found in gardens. The woodlice (page 23) belong to the class, Crustacea, which are usually aquatic and include crabs and lobsters. Although they will drown in water, they will also shrivel up in dry air, which is why they are usually found in damp places. Millipedes (page 22) with a horny outer layer and many pairs of legs are members of the class, Diplopoda. Another class of many-legged arthropoda is the Chilopoda or centipedes (page 21), which move much faster than millipedes and which are carnivores searching for other invertebrates on which to prey. The other two classes of arthropods found in gardens are insects (pages 24–77) and spiders (pages 78–83). Insects go through a number of stages from the egg to the adult animal. Adult insects have bodies made up of a head, which contain the eyes, antennae and mouth-parts; a thorax made up of three segments with a pair of legs on each segment and a pair of wings on the rear two segments; and an abdomen of 11 segments. Spiders have bodies in two segments, on the first of which are four pairs of legs and its mouth-parts and sensory apparatus.

Key

The simple key below will help you to find out where to look when you are trying to identify an invertebrate.

Earthworms and other worms
Segmented worms belong to the phylum Annelida. They have bodies divided into segments of more or less the same width and containing muscle. There are three classes of worms containing about 7000 species worldwide.

Without legs?
Worm-like?
WORMS page 16

Slugs and snails
Land snails and slugs belong to the phylum Mollusca. They are characterised by their soft, unsegmented bodies, which usually have a muscular "foot" and often a shell as well. There are about 90 species in the British Isles. Snails must keep their bodies moist, and as they move they lose moisture through the mucus they exude (the familiar glistening snail trails). In both very cold and very dry weather, they seek shelter among stones and leaf litter and become inactive. Most species are hermaphroditic, with each individual being able to both lay eggs and fertilize another. Food is found by smell and taste.

Thick with shell?
SNAIL page 18

Thick without shell?
SLUGS page 20

Centipedes and symphylans
Centipedes belong to the huge phylum Arthropoda, which includes woodlice, insects and spiders. Equipped with between 20 and 202 legs, depending on the species, they are hunters that live beneath stones, dead wood and leaf litter. They prey upon spiders, woodlice, slugs, worms and insects, using their poison claws to subdue fast-moving prey.

Many legs?
Flat segmented body and fast-moving?
CENTIPEDES page 21

Millipedes
These ground-living arthropods are vegetarians. Those that live in the soil are blind. Some species are serious garden pests, and all are preyed upon by spiders, frogs, toads, small mammals and birds, particularly starlings. Only 50 of the 8,000 or so known millipede species live in the British Isles, and about a quarter are found in gardens.

With round segmented armour and slow-moving?
MILLIPEDES page 22

Woodlice
These small, harmless oval creatures are crustaceans. Their skins are not completely waterproof, will dehydrate very quickly if exposed to dry air. They are found in damp places, spending the daylight hours under cover and emerging at night to feed. The body is divided into three sections: a small head with two pairs of antennae and tough jaws, the body, which is covered by seven overlapping plates with seven pairs of legs on the underside and finally the abdomen.

Rounded body?
WOODLICE AND
PILL-BUGS page 23

Insects
Insects have a head, a thorax and an abdomen. The head has a pair of antennae, mouthparts and eyes. The three-segmented thorax bears three pairs of legs and may have a pair of wings on the second and third segments. The abdomen has 11 segments. The life-cycle of most insects starts with the egg, followed by a larval stage and a pupal stage before the adult emerges.

Six legs?
INSECTS page 24

Spiders and harvestmen
The bodies of spiders and harvestmen are divided into two parts. The first part carries the mouthparts, eyes and four pairs of legs and the second part, the abdomen, is limbless. Spiders are predatory, using several techniques of capture. Harvestmen have very long legs and short compact bodies.

Eight legs?
SPIDERS AND
HARVESTMEN page 78

Earthworms and other worms

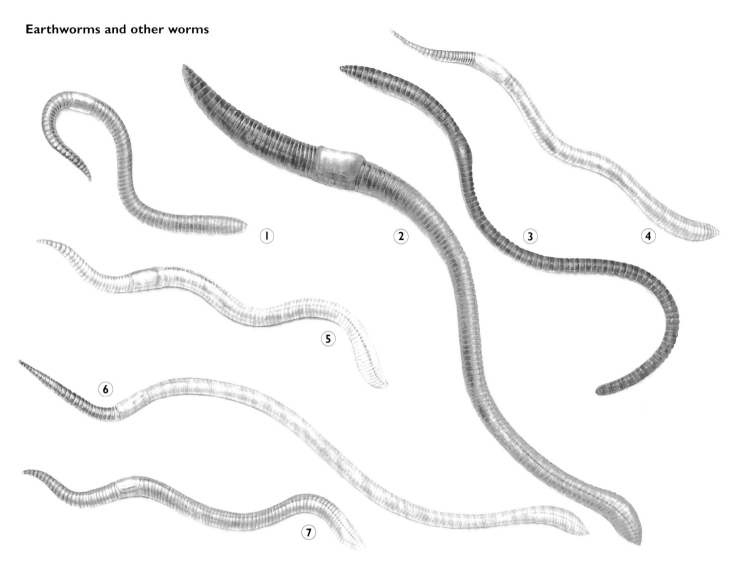

1 Chestnut worm
Lumbricius castaneus
SIZE AND DESCRIPTION 30–70 mm long. Brown with bright orange clitellum on segments 29 to 32.
HABITAT Widespread over Europe, where soil is suitable. Absent from Spain and Portugal.
FOOD/HABITS Behaviour similar to common earthworm.

3 Angler's red worm
Lumbricus rubellus
SIZE AND DESCRIPTION 25–140 mm long. Bright red-brown. Clitellum on segments 28 to 31.
HABITAT Over most of Europe in any soil that is not too wet or too acid.
FOOD/HABITS Feeds and mates in the same way as common earthworm.

5 Blue-grey worm
Octolasion cyaneum
SIZE AND DESCRIPTION 40–180 mm long. Greyish-blue with a red clitellum on segments 29 to 33 or 30 to 34. 150–165 body segments. Yellowish tail.
HABITAT Widespread in moist soils across Europe.
FOOD/HABITS Feeds beneath the soil on decaying material. Emits a thick milky fluid when disturbed.

2 Common earthworm
Lumbricus terrestris
SIZE AND DESCRIPTION 90–300 mm long. Bright pink to reddish-brown, sometimes with a violet tinge. The worm consists of about 150 segments with a reddish-orange "saddle" or clitellum over segments 33 to 36.
HABITAT Over most of Europe in any soil that is not too wet or too acid.
FOOD/HABITS Swallows soil and digests any organic material.

4 Turgid worm
Allolobophora nocturna
SIZE AND DESCRIPTION 90–180 mm long. Dark reddish-brown becoming purplish towards the rear end. 200–246 body segments.
HABITAT Over most of Europe in any soil that is not too wet or too acid. Not as widespread as *A. longa*.
FOOD/HABITS Nocturnal.

6 Long worm
Allolobophora longa
SIZE AND DESCRIPTION 90–170 mm long. Clitellum covers eight or nine segments between segments 27 and 36. Body has 170–190 segments.
HABITAT Widespread over Europe in gardens, cultivated land and woodlands on chalky or loamy soil.
FOOD/HABITS Same as earthworm.

7 *Octolasion lacteum*
SIZE AND DESCRIPTION 25–160 mm long. Bluish with an orange or pink clitellum on segments 30 to 35. 100–135 body segments.
HABITAT Under stones and logs, decaying leaves and compost in pasture, arable and gardens. Widespread in southern and western Europe.
FOOD/HABITS Feeds beneath the soil on decaying material.

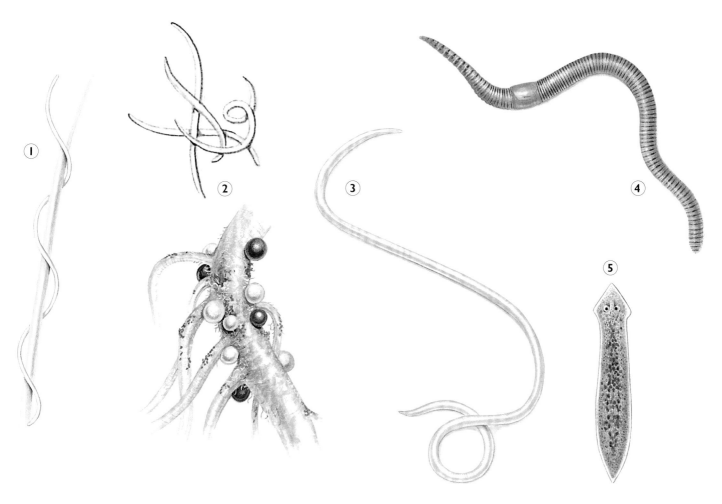

1 Thunderworm
Mermis nigrescens
Size and description
Up to 50 cm long. Looks
like a piece of brown or
white cotton. The body
is not segmented.
Habitat Soils across
Europe, but commoner
in the south.
Food/habits Lives in the
soil, but emerges after
rain and twines around
low-growing plants.
Females lay eggs on
plants, which are
consumed by insects.
Young worms hatch
inside the insect and feed
on its fluids, emerging on
maturity and living in the
soil. The host is
weakened but not
necessarily killed.

**2 Potato root
eelworm**
Globodera rostochensis
Size and description
Up to 30 mm long.
Whitish nematode.
Habitat Farmland and
gardens where potatoes
are grown.
Food/habits Females lay
up to 600 eggs in the soil
in a tough sac that
survive for up to 10 years
before hatching in the
potato roots (as
illustrated above).

3 *Toxocaris canis*
Size and description
Up to 300 mm long.
Round and pale with
no segments.
Habitat Widespread
parasitic roundworm or
nematode carried in the
faeces of cats and dogs.
Food/habits Eggs are laid
in faeces and can be
transferred to humans,
in which it can be
very dangerous.

4 Brandling worm
Eisenia foetida
Size and description
35–30 mm long. Red or
purplish brown with pale
rings on each segment.
Habitat Common in
compost heaps and
under rotting, fallen
tree-trunks over much
of Europe except the
far north.
Food/habits Feeds on
richly organic material
and is used in compost
bin wormery. Emits a
smelly yellow liquid
when handled.

5 *Dugesia tigrina*
Size and description
Can reach up to 30 mm
long but size very
variable. Mottled grey-
brown with paler
underside.
Habitat An American
species frequently
recorded in garden ponds
into which it has been
introduced accidentally
with ornamental fish.
Food/habits Feeds on
a wide range of
invertebrates. It usually
reproduces asexually.

1 Strawberry snail
Trichia striolata
SIZE AND DESCRIPTION Flattened-spiral shell is up to 14 mm in diameter. Colour of shell varies from yellow to reddish-brown or purple, with a prominent white ring around the shell-mouth.
HABITAT Hedgerows, gardens and wasteland with plenty of moisture. British Isles and across Europe to Hungary.
FOOD/HABITS Mainly nocturnal, but also browses on plants after rain. Shelters under plants during the day.

2 Garden snail
Helix aspersa
SIZE AND DESCRIPTION Large, round shell has a diameter of 25–40 mm and a wide, round, white-lipped mouth. Brown or yellowish shell has pale flecking and up to five darker spirals.
HABITAT Parks, woods and wasteland in Europe. Frequently found in gardens, especially in northern regions, where it needs shelter from the winter cold.
FOOD/HABITS Feeds on low-growing plants. It is active at night, and congregates during the day at regular resting places.

3 Rounded snail
Discus rotundatus
SIZE AND DESCRIPTION Shell has a diameter of up to 6 mm. Pale yellow-brown shell, with broad, reddish stripes. Grey body.
HABITAT Found everywhere, except in the driest habitats. Very common in leaf litter and in garden compost heaps.
FOOD/HABITS Feeds on decaying plant material and fungi.

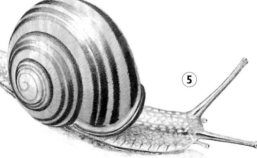

4 White-lipped snail
Cepaea hortensis
SIZE AND DESCRIPTION Shell is 14 x 17 mm, with a lip that is usually white, sometimes brown. Smaller than the brown-lipped snail. Shell has up to five dark spirals, but may have none.
HABITAT Woods, hedges and gardens, especially in moist habitats. Found throughout Europe, as far north as Iceland.
FOOD/HABITS Similar to those of the brown-lipped snail.

5 Brown-lipped snail
Cepaea nemoralis
SIZE AND DESCRIPTION Shell is 18 x 22 mm, with a lip that is usually brown, sometimes very pale. Shell colour varies from straw to yellow to pink to brown. Up to five dark spiral bands, but sometimes none.
HABITAT Woods, hedges, rough vegetation and gardens (but less frequently than the white-lipped snail). Found across Europe, but not as far north as the white-lipped snail.
FOOD/HABITS Eats grass and low-growing plants. Feeds at night and after rain, often alongside other species.

6 Kentish snail
Monacha cantiana
SIZE AND DESCRIPTION Shell is up to 20 mm in diameter. Shell colour varies from off-white to reddish-pink, often darker near the mouth.
HABITAT Long grass in hedgebanks, wasteland, fields and herbaceous garden borders on calcareous soils. Widespread in south and central Europe.
FOOD/HABITS Feeds on decaying vegetation, including lawn cuttings.

1 Shelled slug
Testacella haliotidea
SIZE AND DESCRIPTION
Up to 120 mm long.
Creamy-white or pale
yellow body, with a small
flat shell at the rear end.
Dark lines run forward at
angles along the sides of
the body.
HABITAT Well-manured
and well-drained soil in
parks and gardens across
western Europe.
FOOD/HABITS Earthworms
are the slug's main food.
It can extend its body to
become narrow enough
to follow worms down
their holes.

2 Door snail
Clausilia bidentata
SIZE AND DESCRIPTION
Brown shell is 12 mm
long and spirals towards
the left.
HABITAT Widespread in
hedgerows, as well as on
rocks and old walls.
FOOD/HABITS Hides in leaf
litter by day and emerges
at night to feed on
lichens and algae.

3 Great pond snail
Lymnaea stagnalis
SIZE AND DESCRIPTION
30 mm long. Yellowish to
dark brown, with a
pointed spiral shell.
HABITAT Large, calcium-
rich ponds and slow-
flowing rivers and canals.
FOOD/HABITS Feeds on
algae and decaying
vegetation. Eggs are laid
in a sausage-shaped
gelatinous sac on the
underside of leaves.

**4 Great ram's-horn
snail**
Planorbis corneus
SIZE AND DESCRIPTION Dark
brown shell is
25 mm in diameter. The
shell's shape gives the
snail its name.
HABITAT Ponds, lakes and
slow-flowing rivers. Also
found in garden ponds,
because it is sold by
aquarium dealers.
FOOD/HABITS Feeds on
algae on stones and
plants. Its eggs, which are
laid on stones, may be
spread to other ponds
when they stick to the
feet of birds.

5 Garlic snail
Oxychilus alliarius
SIZE AND DESCRIPTION
Dark brown shell is
6 mm across and very
glossy. Body is black.
HABITAT Found in leaf-
litter in a range of
habitats. Lurks under
stones and in garden
compost-heaps.
FOOD/HABITS Feeds
mainly at night on fungi
and rotting vegetation
on the ground, but also
climbs walls and trees
on damp nights.

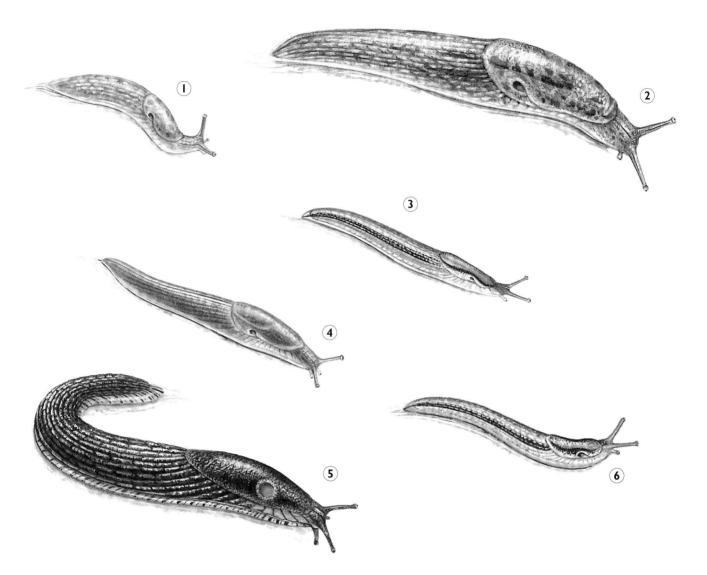

1 Netted slug
Deroceras reticulatum
SIZE AND DESCRIPTION Small slug, up to 50 mm long. Light brown or dark grey, with darker flecks and rectangular tubercles that create a netted pattern. Short keel at rear.
HABITAT Gardens, hedges, arable fields and rough pasture. One of Europe's commonest and most widespread slugs.
FOOD/HABITS Exudes white mucus when disturbed. Eats a wide range of plants, especially newly planted seedlings, and is a garden pest.

4 Smooth jet slug
Milax gagates
SIZE AND DESCRIPTION Up to 75 mm long. Greyish brown and heavily speckled, with a keel darker than the rest of the body. Exudes clear mucus. Very similar to Sowerby's slug, *Milax sowerbyi*, which has a yellowish or orange keel and exudes yellowish mucus. Relatively dry-skinned.
HABITAT Gardens and arable fields. Most common in western Europe.
FOOD/HABITS Feeds on roots and tubers.

2 Great grey slug
Limax maximus
SIZE AND DESCRIPTION Up to 200 mm long. Pale grey, heavily marked with dark spots, appearing striped at the end of the body. Short keel on rear end of the body.
HABITAT Woods, hedges and gardens – especially around compost heaps – over much of Europe, except in the far north.
FOOD/HABITS Eats fungi and rotting plant material. Mating involves two individuals climbing a fence, tree-trunk or wall, and then lowering themselves on a string of mucus. Each of these hermaphrodites then lays eggs.

5 Large black slug
Arion ater
SIZE AND DESCRIPTION Up to 150 mm, but may reach 200 mm when extended. Colour ranges from jet-black through orange to creamy-white with an orange fringe. Back is covered with elongated tubercles. No keel. Sticky mucus.
HABITAT Well-vegetated habitats across Europe to Iceland. Darker forms are most common in the north, paler ones in the south.
FOOD/HABITS Nocturnal feeder on dung, plants and carrion. Will eat grass cuttings after rain.

3 Bourguignat's slug
Arion fasciatus
SIZE AND DESCRIPTION Up to 40 mm long. Body is grey, with dark patches. Similar to the garden slug, but with a white underside.
HABITAT Gardens and woods.
FOOD/HABITS Feeds on fungi and decaying material.

6 Garden slug
Arion hortensis
SIZE AND DESCRIPTION Up to 40 mm long. Bluish-black and paler on the flanks, with an orange underside. Mucus is orange or yellow.
HABITAT Most common on cultivated land, but can also be found in woods and gardens. All Europe, except the far north.
FOOD/HABITS Eats any plants near the ground, and is a serious pest of strawberries, lettuces and seedlings.

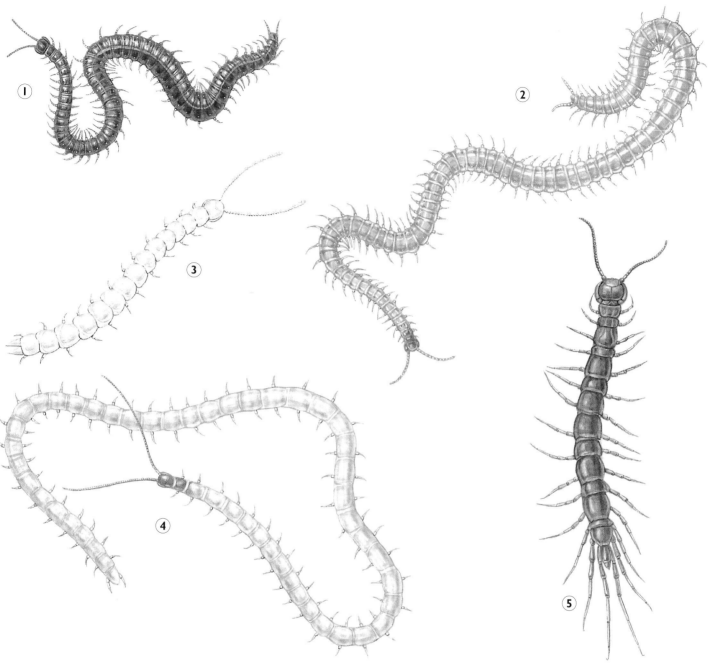

1 *Geophilus carpophagus*
SIZE AND DESCRIPTION
About 40 mm long, but
may be longer, and up to
1.5 mm wide. Reddish
brown. Very flexible body,
with 45–55 pairs of legs.
HABITAT Woodlands,
orchards and gardens
across Europe. Also found
in cellars and damp
outbuildings.
FOOD/HABITS Soil-
dwelling, fast-moving
predator.

2 *Haplophilus subterraneus*
SIZE AND DESCRIPTION
Up to 70 mm long. Yellow
or pale colour. Tapering
head, typical of burrowing
centipedes. It may glow
when disturbed at night.
Very flexible body, with
77–83 pairs of legs.
HABITAT Common in
grassland and arable
fields, but also found in
gardens across most of
Europe. Lives under
stones and leaf litter.
FOOD/HABITS Nibbles
plant roots and eats small
subterranean animals.

3 Garden centipede
Scutigerella immaculata
SIZE AND DESCRIPTION
About 7 mm long, with
12 pairs of short legs and
long antennae. This is not
a centipede, but part of
another arthropod group
called symphylans.
HABITAT Soil and leaf
litter in several habitats,
including woodlands and
gardens.
FOOD/HABITS A
symphylands, not a
centipede, it feeds
on plant material,
especially seedlings.

4 *Necrophloeophagus
longicornis*
SIZE AND DESCRIPTION
About 30 mm long, but
may be longer, and 1 mm
wide. Yellow, with long
antennae. Very flexible
body, with 49–51 pairs
of legs.
HABITAT Ranges from
alpine terrain to
seashores throughout
Europe, and is abundant
in gardens.
FOOD/HABITS Burrowing
predator.

5 Common centipede
Lithobius forficatus
SIZE AND DESCRIPTION
18–30 mm long and 4 mm
wide. Shiny chestnut-
brown. Adults have 15
pairs of legs; hatchlings
have seven pairs, growing
extra pairs at each moult.
Head is rounded.
HABITAT Widespread
throughout Europe, from
moorlands to coasts.
Abundant in gardens.
FOOD/HABITS Hides under
stones and logs in
daytime. At night, it hunts
insects, worms, slugs and
other centipedes.

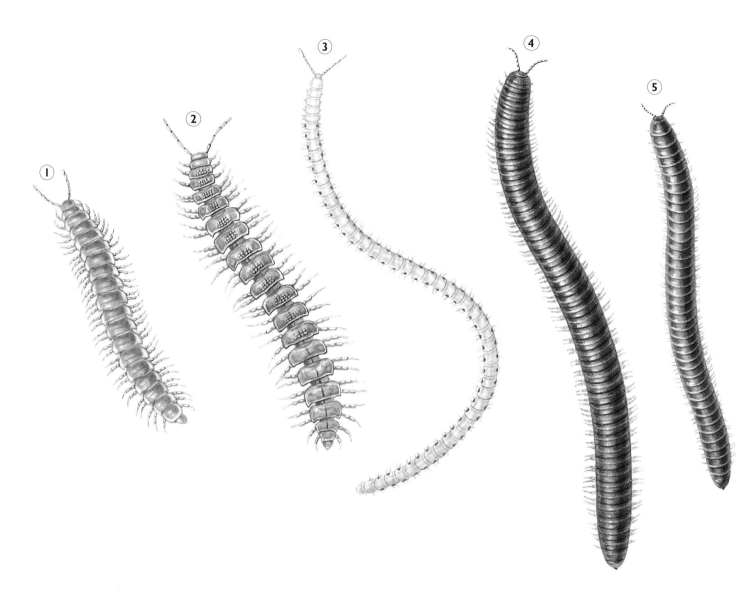

1 Greenhouse millipede
Oxidus gracilis
SIZE AND DESCRIPTION
Up to 23 mm long and 2.5 mm wide. Resembles the flat-backed millipede, but *Oxidus gracilis* has a smoother, more rounded back. Two pairs of legs on each body segment.
HABITAT A tropical species that has become established in European greenhouses.
FOOD/HABITS Feeds on decayed and living plant matter. Lays eggs throughout the year.

2 Flat-backed millipede
Polydesmus angustus
SIZE AND DESCRIPTION
Up to 25 mm long and 4 mm wide. Looks like a centipede, but it has two pairs of legs on each segment. It has flattened body segments and 37 pairs of legs.
HABITAT Leaf litter, turf, organically rich soils and garden compost heaps in most of Europe.
FOOD/HABITS Eats decaying vegetation, but also nibbles plant roots and soft fruit, including strawberries.

3 Spotted snake millipede
Blaniulus guttulatus
SIZE AND DESCRIPTION
About 15 mm long and no more than 0.7 mm in diameter. Pale, with red spots on either flank. The spots are glands that release a repellent liquid as a defence against predators.
HABITAT Arable fields and gardens on heavy, damp soils across Europe.
FOOD/HABITS Burrows and eats rotting material, but feeds on roots in dry weather.

4 Snake millipede
Tachypodiulus niger
SIZE AND DESCRIPTION
Up to 50 mm long and 4 mm in diameter, with a cylindrical, shiny black-brown body that tapers at each end.
HABITAT Hedges, garden borders and woodlands across most of Europe. Lives in the surface layer of soil, under loose bark and amongst leaf litter.
FOOD/HABITS Feeds at night on living and decaying plants. Climbs raspberry canes to reach fruit. Coils up when disturbed.

5 *Cylindroiulus londinensis*
SIZE AND DESCRIPTION
Up to 50 mm long and 4 mm in diameter along the entire length of the shiny black body. (The body of the similar-looking snake millipede tapers slightly towards each end.)
HABITAT Hedges, woodlands and gardens across Europe. Found in the surface layer of soil, loose bark and leaves.
FOOD/HABITS Feeds on plant material. Nocturnal.

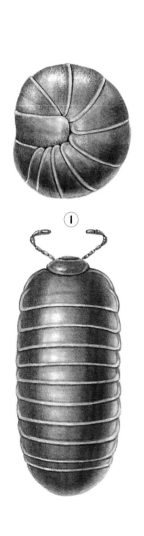

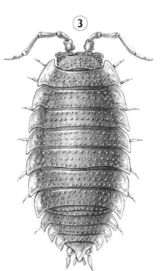

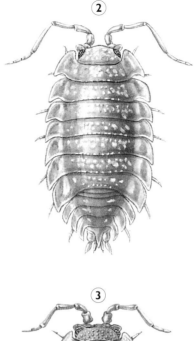

1 Pill millipede
Glomeris marginata
SIZE AND DESCRIPTION Up to 20 mm long and 3 mm wide, with 17–19 pairs of legs. Often confused with the pill woodlouse. They look similar but its dorsal plates are shinier and deeper, and there is a broad, almost semi-circular plate at the rear.
HABITAT Leaf litter and turf in woodlands, hedges and gardens across Europe. Able to endure drier conditions than other millipedes.
FOOD/HABITS Eats stems and dead vegetation. Rolls up into a ball when disturbed.

2 Common woodlouse
Oniscus asellus
SIZE AND DESCRIPTION About 15 mm long and 8 mm wide. The body sections are not obvious, giving the woodlouse a smooth outline. Shiny grey in colour, with yellow or cream blotches and pale edges to the plates on the back.
Looks flatter than other woodlice.
HABITAT Abundant on lime-rich soils across Europe, especially in gardens under logs and in compost heaps. One of the commonest garden species.
FOOD/HABITS Very fond of rotting wood and other plants.

3 *Porcellio scaber*
SIZE AND DESCRIPTION About 17 mm long. Dull grey in colour, with noticeable tubercles and pale spots.
HABITAT Very common throughout Europe. Able to tolerate drier conditions than *Oniscus asellus*, with which it is often found beneath logs or under stones.
FOOD/HABITS Feeds at night on algae on walls and tree-trunks.

4 *Androniscus dentiger*
SIZE AND DESCRIPTION About 6 mm long. Body is pale, sometimes pinkish, usually with a darker central stripe. Prominent, spined tubercles cover the body.
HABITAT Found throughout Europe, except in the far north. Lives in compost heaps, leaf litter and cellars.
FOOD/HABITS Is attracted by limestone on walls.

5 Pill woodlouse
Armadillidium vulgare
SIZE AND DESCRIPTION 18 mm long. The smooth, shiny slate-grey body has a domed back with a tinge of blue, or sometimes brown or yellow.
HABITAT Dry grasslands, but restricted mainly to lime-rich soils. More tolerant of drier habitats than other species. Widely distributed and common over much of Europe, except in the north. Often found at the base of walls.
FOOD/HABITS Rolls itself into a ball, or "pill", if disturbed. When curled up, it can be distinguished from the pill millipede by the numerous small plates at its rear.

Insects

This is a huge class of animals with a million or so species already identified and more remain to be described. Almost 100,000 species of insect are found in Europe. Some are so small that a microscope is needed to see them clearly, while some of the moths and dragonflies have wingspans of up to 12 cm. The form which they take is also very varied, but they do share certain anatomical characteristics. The bodies of adult insects have three main parts: the head, the thorax and the abdomen.

The head has a pair of compound eyes, whose surfaces are faceted with tiny lenses. The number of these lenses or facets varies, but dragonflies, which are swift fliers and active predators, have several thousand in each eye while some soil-dwelling insects may have none. In addition some insects have ocelli, very simple eyes on the front of the head and probably having the function of detecting light rather than producing images. There are two antennae, which are the sensors of smell and touch. Some species have simple antennae that are a series of similar segments well-supplied with nerve endings. In other species the antennae may be more complex. The head also contains the mouth-parts which are complex and which vary according to the feeding methods of each species. The basis of the mouth-parts are a pair of jaws, a pair of secondary jaws and a lower lip. There are also four palps which examine the food before it is eaten. The secondary jaws and the lower lip hold the food steady while the jaws cut it up. The mouth-parts of species which feed on liquids have been modified quite dramatically. True bugs which feed on the sap of plants have piercing mouth-parts. Mosquitoes and horse-flies have long, needle-like jaws with which they pierce skin and withdraw blood. Moths and butterflies have no jaws, but the secondary jaws have become linked together to form a long proboscis through which they can suck nectar.

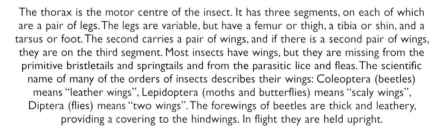

The thorax is the motor centre of the insect. It has three segments, on each of which are a pair of legs. The legs are variable, but have a femur or thigh, a tibia or shin, and a tarsus or foot. The second carries a pair of wings, and if there is a second pair of wings, they are on the third segment. Most insects have wings, but they are missing from the primitive bristletails and springtails and from the parasitic lice and fleas. The scientific name of many of the orders of insects describes their wings: Coleoptera (beetles) means "leather wings", Lepidoptera (moths and butterflies) means "scaly wings", Diptera (flies) means "two wings". The forewings of beetles are thick and leathery, providing a covering to the hindwings. In flight they are held upright.

The abdomen is the centre of digestion and excretion. It is also where the sexual organs are situated. Most insects have eleven abdomenal segments. At the tip there are a pair of cerci or tail-feelers. The male often has a pair of claspers on the ninth segment: these are used for holding the female while mating. She has an ovipositor between the eighth and ninth segments. This is long in the cases of ichneumons and the greater horntail. It has lost its egg-laying function among bees and wasps, where it has become a sting.

Insects go through a series of developmental stages, or metamorphoses. The nature and timing of these stages differ between groups and species. The first stage is the egg. The hatchling from the egg looks nothing like its parent except in the case of some of the most primitive insects. The best known life-cycle is that of butterflies, whose larvae are caterpillars, which then pupate to become chrysalids, from which the adults emerge. The caterpillar's sole purpose is eating and as it grows it sheds its outer covering in a series of moults. As an insect moults it becomes slow and seeks cover because it is very vulnerable, which means the process of moulting is difficult to see in the wild. Dragonflies lay their eggs in water and the larvae develop underwater, emerging up the stems of plants and then shedding their larval casing to emerge as adult insects. Grasshoppers have a partial metamorphosis with the young or nymphs, looking like tiny adults and growing in a series of nymphal stages, shedding their skins several times until they reach adulthood.

Key

**The simple key
below will help you to find out where to look when you are trying
to identify an insect.**

Grasshoppers and crickets
There are about 250 species in Europe (30 in UK), all have strong
hindlegs for jumping. Some have wings. They produce a greater variety
of sounds than any other order of insects. Depending on the species,
the sounds are created by friction between the legs and wings, the
wings and the body, or by rubbing the mandibles together.

**Without wings?
Bullet body with well-
developed hindlegs?**
GRASSHOPPERS AND CRICKETS
page 31

Bristle-tails, two-pronged bristle-tails and springtails
The most primitive of insects. The bristle-tails have three "tails", each
fringed with minute bristles, and tapering, wedge-shaped bodies. Two-
pronged bristle-tails have only two tails and three clear segments to
the thorax. The bodies of springtails have fewer segments than other
insects. Springtails possess appendages beneath the abdomen that
enable them to spring forward.

**With long, slender
hindtails?
Three tails (cerci)?**
BRISTLETAILS page 27
Two tails (cerci)?
TWO-TAILED BRISTLETAILS
page 27

Earwigs
Of the 1,300 known species of earwig, 34 occur in Europe and four
in the British Isles. Earwigs are elongated insects with pincer-like
cerci, which are longer and more curved in males. They are nocturnal,
ground-living scavengers that eat both plant and animal matter. They
hibernate during winter months.

**With pincers?
Brown insects, often
under stones or leaf-litter?**
EARWIGS page 32

Cockroaches
These flat, fast-running insects have long antennae and bristly legs.
There are about 3,500 species of cockroach, most of which occur in
the tropics and sub-tropics. European species are small, but several
non-native species have been accidentally introduced into Europe,
and typically live in warm buildings.

**With short tails or none
at all?
Flattened insects with
spiky legs and a broad
pronotum almost covering
head?**
COCKROACHES page 32

Aphids
These sap-sucking bugs smother young shoots of many plants in
summer. The excess sugar is exuded as "honeydew" which is eaten
by ants. Responsible for the sticky deposits found on some leaves.

**Small with bodies much
wider than head?
On plants?**
APHIDS page 36

Fleas
Fleas are tiny, wingless insects that suck the blood of birds and
mammals. Only 5% of the 1,400 known species of flea are parasites
of birds. Species of fleas often breed on one host species, or a group
of similar hosts. The bodies are flattened to enable them to move
through fur or feathers.

**Tiny with rounded body?
On other animals?**
FLEAS page 37

Psocids
Psocids are very small winged and wingless insects. They are also
known as booklice, barklice and dustlice. So far, about 2,000 species
worldwide have been described. All have biting jaws.

**Tiny insect with wings held
like a tent and longer than
abdomen? Flattened with
broad head?
Commonly found indoors?**
PSOCIDS page 37

Butterflies and moths
Known as Lepidoptera, with over 100,000 species worldwide, with
2,300 in the British Isles (about 70 are butterflies). Sizes vary from
tiny to large. Mouthparts adapted to suck nectar through a proboscis.
Four wings tend to be large and flattened. Butterflies can be
distinguished from moths by their clubbed or knobbed antennae. At
rest, butterflies tend to hold their wings vertically above the body.

**Large wings covered with
fine scales and often
patterned? Wings held
upright at rest?**
BUTTERFLIES page 38
Wings held flat at rest?
MOTHS page 42

Key

Please note that beetles and some insects look at first glance as if they do not have wings.

Scorpion flies
These winged insects have distinctive "beaks". About 300 species are known, of which only four are found in the British Isles.

One pair of wings? Round, translucent forewings, hindwings barely noticeable?
FLIES   page 53

Thrips
These tiny, usually dark-coloured insects have two pairs of feathery wings. They are known as thunder-flies, because of their habit of flying in still, thundery weather. Females have curved ovipositors. Over 3,000 species are known.

Two pairs of wings? Feathery wings? Tiny insect?
THRIPS   page 37

Dragonflies and damselflies
About 120 species of dragonfly and damselfly breed in Europe, 38 of them in the UK. They have large eyes that cover most of the head, a thorax with four wings and six legs, and a long body with ten segments. Dragonflies rest with their wings at right angles to the thorax, while damselflies hold their wings closed over the abdomen.

Forewings and hindwings similar size? Translucent wings with fine veins?
DRAGONFLIES   page 28
LACEWINGS   page 37

Lacewings
A group of soft-bodied insects, the lacewings are characterised by large, flimsy wings. There are about 4,500 known species, of which about 60 are British species in the UK.

Whiteflies
Tiny homopterans, with a wingspan of 3 mm. Resembling minute moths, usually found on the underside of leaves.

Tiny insect covered in fine white powder?
WHITEFLIES   page 36

Bees
Bees may be either social or solitary, depending on the species. They all feed and rear their young on pollen and nectar. Bodies are rather hairy and pollen is often carried back to the nest on the hairy legs. A social species, living in colonies with queens and workers. Bees build their nests from wax from their own bodies. Although honey bees sting in defence of their colonies, other bees will sting if handled.

Membranous wings with forewings larger than hindwings? With distinct "waist"?
WASPS, BEES, ICHNEUMONS AND ANTS page 60

Bugs
There are about 75,000 species in this order of insects. Europe has some 8,000 species, of which 1,700 are found in Britain and Ireland. Shapes and sizes vary, but they all share a piercing beak, like a tiny needle, which is used to extract juices from plants and other animals.

Hard forewings covering all or most of abdomen?
BUGS   page 33,
BEETLES   page 70

Beetles
Largest order of insects called coleoptera, which means "leather wings". The leathery casing on the abdomen are the forewings, held vertically when it flies. Clumsy and distinctive in flight, but usually seen on the ground, under stones and logs.

Leathery or partly leathery wings? Small wings which do not cover most of abdomen?
EARWIGS   page 32
BEETLES   page 70

Mantids
Related to cockroaches, these predatory insects have leathery forewings and a long thorax. There are about 2,000 known species, of which 18 are found in southern Europe.

Enlarged front legs? Wings resting along abdomen?
MANTIDS   page 32

Sawflies
Although they belong to the same order as bees, wasps and ants, sawflies have no "waist". Many species have saw-like ovipositors with which females cut slits in plants before laying their eggs. Larvae look like moth caterpillars, but they have at least six pairs of legs (compared with moth caterpillars' five pairs).

Without "waist"
SAWFLIES   page 62

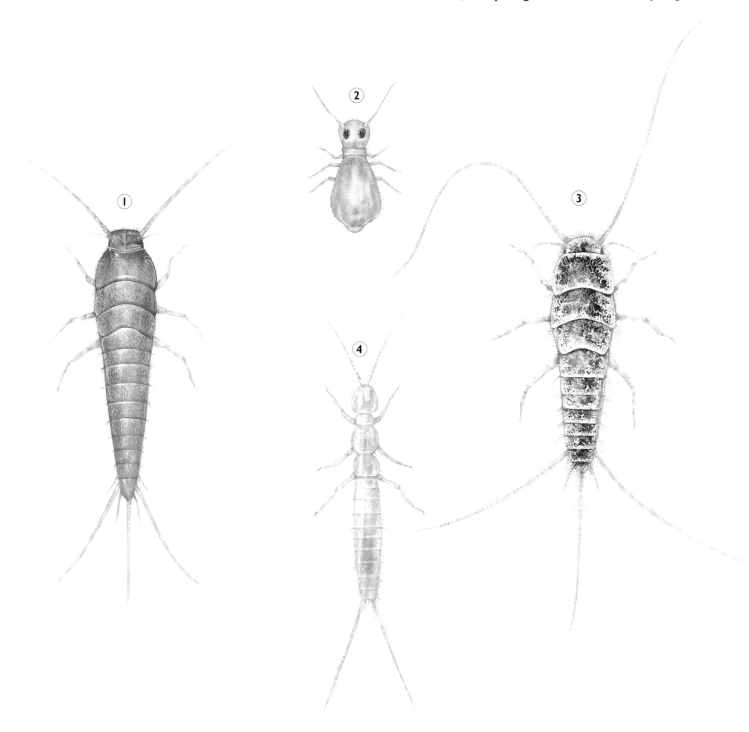

1 Silverfish
Lepisma saccharina
SIZE AND DESCRIPTION
13 mm long, with three long, tail-like appendages. Tapering body is covered in shiny, silvery scales. Fine antennae and small eyes.
HABITAT Houses and garden sheds, preferring damp places.
FOOD/HABITS Nocturnal feeder that eats starchy material, such as flour, paper and gum.

2 Lucerne flea
Sminthurus viridis
SIZE AND DESCRIPTION
1.5 mm long. Yellowish-green, with a round abdomen. Antennae are longer than the head.
HABITAT Meadows and arable fields.
FOOD/HABITS Feeds by nibbling the stems and leaves of clover and pea plants.

3 Firebrat
Thermobia domestica
SIZE AND DESCRIPTION
13 mm long, with three long, bristly tails. The body is brown and tapering. Antennae are longer than the silverfish's.
HABITAT Found indoors, in warm places such as heating ducts. Often found in bakeries.
FOOD/HABITS Feeds on starchy material, such as flour.

4 Two-tailed bristletail
Campodea fragilis
SIZE AND DESCRIPTION
10 mm long, with two bristly, tail-appendages. Whitish body. Thorax has three clearly separated segments. No eyes. Several similar species.
HABITAT Abundant in compost heaps and decaying vegetation.
FOOD/HABITS Carnivore and scavenger.

Damselflies

1 Blue-tailed damselfly
Ischnura elegans
SIZE AND DESCRIPTION 31 mm long, with a 30–40 mm wingspan. Dark bronze/black abdomen. Light-blue band on the 8th segment. Steady flier.
HABITAT Lowland pools and slow-flowing rivers. Usually the first damselfly to visit newly dug garden ponds. Absent from Iceland, Spain and most of Scandinavia.
FOOD/HABITS Flies early May to early September. Larvae favour midge and mayfly larvae.

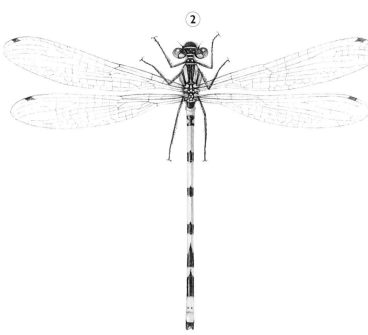

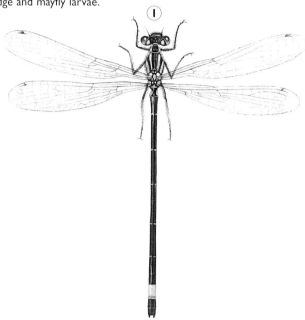

3 Large red damselfly
Pyrrhosoma nymphula
SIZE AND DESCRIPTION 36 mm long, with a 38–48 mm wingspan. Red abdomen, with black markings from the 7th to 9th segments.
HABITAT Clear streams, ponds, lakes and canals across Europe, except northern Scandinavia, Iceland and Sardinia.
FOOD/HABITS Flies late April to late September. Usually in large numbers. Rests on marginal plants. Feeds on small insects.

2 Common blue damselfly
Enallagama cyathigerum
SIZE AND DESCRIPTION 32 mm long, with a 36–42 mm wingspan. Male has a blue abdomen with black spots. The 8th and 9th segments are all blue. Female has a yellowish or bluish abdomen, with variable dark markings. Strong flier.
HABITAT Pools, ponds, peat bogs and lakes. Found throughout Europe. Absent in Iceland, and much of the Mediterranean,
FOOD/HABITS Flies mid-May to mid-September. May swarm in large numbers over water. Will pounce on dark spots on leaves, mistaking them for aphids.

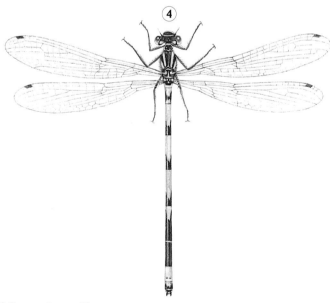

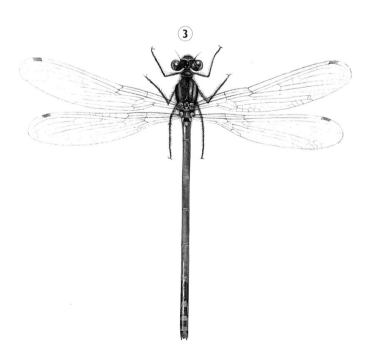

4 Azure damselfly
Coenagrion puella
SIZE AND DESCRIPTION 33 mm long, with a 36–44 mm wingspan. Male has a blue abdomen, with black markings and a completely blue 8th segment. Female has a dark abdomen, with blue or green markings.
HABITAT Pools and lakes up to 1,800 m. Found from Ireland and southern Scotland across Europe, and south to North Africa.
FOOD/HABITS Flies mid-May to late August. Often seen in sunny meadows. Larvae favour small crustaceans and midge larvae.

1 Southern hawker

Aeshna cyanea

SIZE AND DESCRIPTION 70 mm long, with a wingspan of 98 mm. Male is brown, with pairs of green markings on the first seven abdominal segments and blue markings on the 8th, 9th and 10th segments. Female has green markings on her brown abdomen.

HABITAT Ponds, pools and lakes up to 1,400 m and slow-flowing rivers. Absent from Iceland, Ireland and northern Scandinavia.

FOOD/HABITS Flies mid-June to October. Males fly at human waist height and are inquisitive. Will attack other hawker dragonflies.

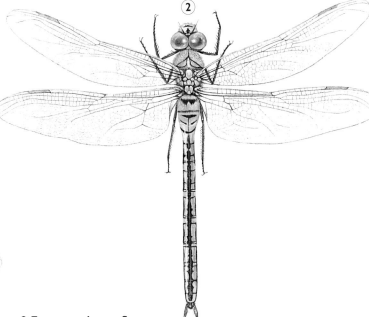

2 Emperor dragonfly

Anax imperator

SIZE AND DESCRIPTION A large dragonfly, 78 mm long, with a wingspan of 107 mm. The thorax is green, and the bright blue abdomen has a thick black stripe down the back. Females are usually green, but may be blue. Males fly strongly, patrolling territory above human head height.

HABITAT Pools, ponds, ditches and slow-flowing rivers across Europe, southwards from Denmark. Known to visit woodland rides and glades when hunting.

FOOD/HABITS Flies late May to mid-August. Hunts flies, moths and beetles, and will even take tadpoles from the water's surface.

3 Common hawker

Aeshna juncea

SIZE AND DESCRIPTION 74 mm long, with a wingspan of 95 mm. Male has a black abdomen, with pairs of blue spots and small yellow marks. Female is brown with yellow marks. Flies well and strongly.

HABITAT Lakes, ponds, peat bogs and still pools. Found south from northern Norway to the Pyrenees. Absent from Iceland. Occurs between 800 m and 1,000 m in southern part of range.

FOOD/HABITS Flies late June to October. Hawks for other insects, often some way from water.

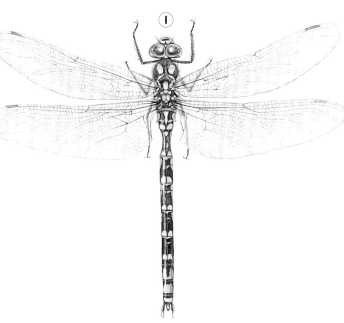

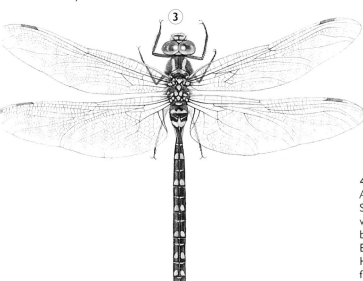

4 Brown hawker

Aeshna grandis

SIZE AND DESCRIPTION 73 mm long, with a wingspan of 102 mm. Brown wings make this species unmistakable. Male has a brown abdomen with bright blue spots. Female has yellow markings on her brown abdomen. Both sexes have diagonal marks on the side of the thorax. Strong flier.

HABITAT Ponds, lakes, canals, peat bogs and slow-flowing rivers. Absent from Iceland, Iberia, Italy, Greece, Scotland and northern Scandinavia.

FOOD/HABITS Flies mid-June to mid-October. Hunts flies, mosquitoes, moths and butterflies.

Dragonflies

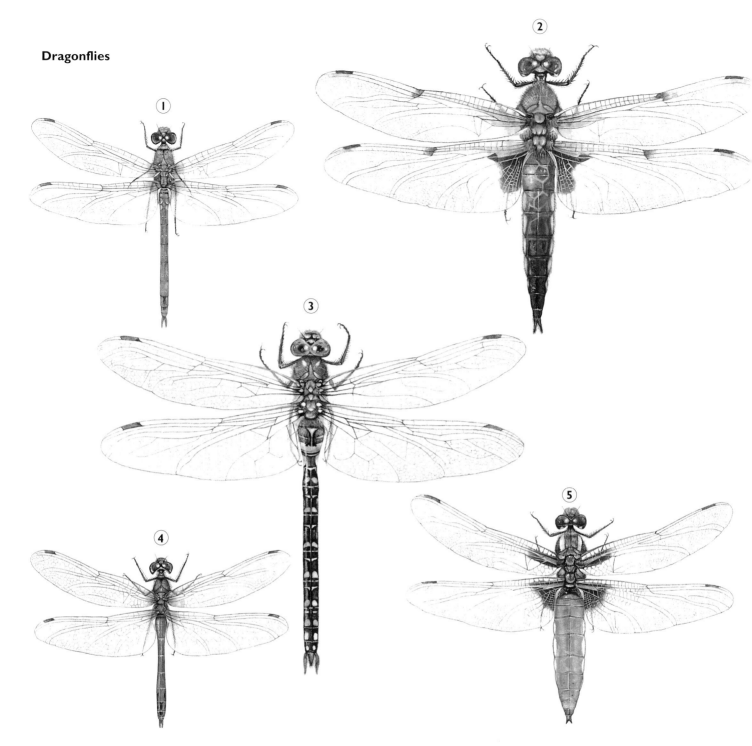

1 Common darter
Sympetrum striolatum
SIZE AND DESCRIPTION
37 mm long, wingspan of
57 mm. Males are red,
with a narrow pointed
abdomen. Females are
greenish yellow. Flies busily.
HABITAT Ponds, lakes,
ditches and brackish
waters up to 1,800 m.
Found across Europe from
Ireland, and south from
southern Scandinavia to
North Africa.
FOOD/HABITS Flies June to
October. Usually seen in
large numbers. Often
perches on twigs.

2 Four-spotted chaser
Libellula quadrimaculata
SIZE AND DESCRIPTION
43 mm long, wingspan of
76 mm. Broad brown
body, with yellow patches
along each side. There are
two dark marks on the
leading edge of each wing.
HABITAT Still water with
plenty of vegetation.
Found throughout
Europe, except Iceland.
FOOD/HABITS Flies mid-
May to mid-August.
Frequently perches in the
open and flies out over
the water. Aggressive and
territorial nature.

3 Migrant hawker
Aeshna mixta
SIZE AND DESCRIPTION
63 mm long, wingspan of
87 mm. Male, dark brown
and blue, has a bright blue
spot on the side at the
base of the abdomen.
Brown female has small
yellow spots. Neat, elegant,
sometimes jerky flight.
HABITAT Still or slow-
flowing water. From
England and Wales across
Europe, south from the
Baltic to North Africa.
Adults migrate.
FOOD/HABITS Flies July to
October. Approachable.

4 Ruddy darter
Sympetrum sanguineum
SIZE AND DESCRIPTION
34 mm long, wingspan of
55 mm. Males are blood-
red, with the tip of the
abdomen club-shaped,
rather than tapering.
Yellow females have black
thorax markings. Flitting,
sometimes jerky flight.
HABITAT Well-vegetated
(even brackish or acid)
pools up to 1,000 m.
Found from Ireland across
Europe, and south from
southern Scandinavia.
FOOD/HABITS Flies June to
October. Often perches.

5 Broad-bodied chaser
Libellula depressa
SIZE AND DESCRIPTION
44 mm long, wingspan of
76 mm. Male has flattened,
fat, pale blue body, with
yellow patches along each
side. Brownish-yellow
females have yellow spots
along each side. Fast flier.
HABITAT Still or slow-
flowing water to 1,200 m.
Across Europe from Wales
and England, and from
southern Sweden south
to the Mediterranean.
FOOD/HABITS Flies early
May to early August. Rests
on waterside plants.

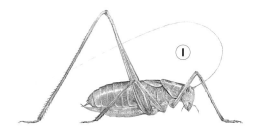

1 Speckled bushcricket
Leptophyes punctatissima
Size and description Body length: 10–14 mm (m); 12–17 mm (f). Yellow-green, with fine red speckles. Male's abdomen has a narrow brown stripe along the top. Wings are short. Female has a long, sickle-shaped ovipositor. Song is a sequence of soft "zb" sounds, at 3–6 second intervals.
Habitat Gardens, parks and forest edges in undergrowth. Widespread from southern Scandinavia to the Mediterranean, including the British Isles.
Food/habits Adults are seen from July or August to October. The speckled bushcricket feeds on leaves, including those of raspberry and rose bushes.

3 Oak bushcricket
Meconema thalassinum
Size and description Body length: 12–15 mm. Pale green, with wings extending beyond the tip of the abdomen. Female has a long, upward-curving ovipositor. Male has two thin, inward-curving cerci, about 3 mm long. Long yellow mark down the back, with two brown flecks on either side.
Habitat Lives in trees, particularly oaks, but may also be found in gardens. Distributed from southern Sweden to northern Spain, Italy and the Balkans.
Food/habits Adults are seen from July to October.

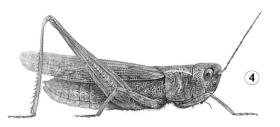

4 Common field grasshopper
Chorthippus brunneus
Size and description Body length: 14–18 mm (m); 19–25 mm (f). Colour can be grey, green, purple or black. Wings are narrow, and extend beyond the tip of the abdomen. Male's abdomen has a reddish tip, a feature that sometimes also occurs in the female. Song is a hard "sst" sound, lasting about 0.2 seconds and repeated at 2-second intervals.
Habitat Widespread in dry, grassy habitats, from Scandinavia to the Pyrenees and Italy. Particularly common in southern England.
Food/habits Adults are seen from July to October.

2 House cricket
Acheta domesticus
Size and description Body length: 16–20 cm. Straw-coloured to brown body, with black marks on the head. Wings extend beyond the tip of the abdomen. Female has a straight ovipositor, up to 15 mm long.
Habitat The house cricket is a native insect of Asia and Africa, but is now widespread in Europe. It lives in buildings, but may also be found at refuse tips in summer. Song is a soft warble delivered at dusk or at night.
Food/habits Feeds on refuse, but will also eat stored food.

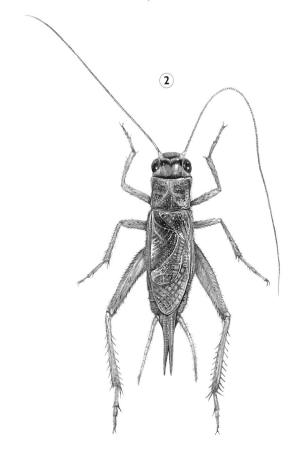

5 Meadow grasshopper
Chorthippus parallelus
Size and description Body length: 13–16 mm (m); 17–33 mm (f). Colour may be grey, green, brown or purple. Wings are short, reaching almost to the tip of the male's abdomen; in the female they are even shorter, being only half as long. Song consists of short, sewing-machine-like chirps in 1-second bursts, followed by 3-second intervals.
Habitat Northern Europe into Scandinavia, where it is found in meadows and grassland that are neither too dry nor too damp. Absent from Ireland and Isle of Man. In the Mediterranean, it tends to occur in mountain regions.
Food/habits Adults are seen from June to November.

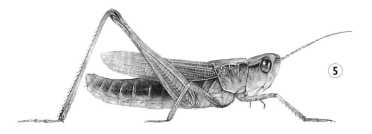

Cockroaches, mantids and earwigs

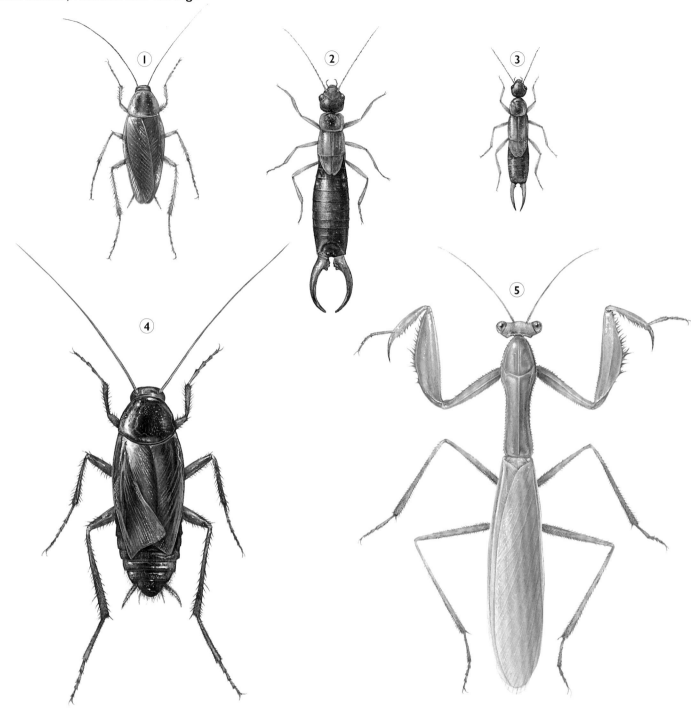

1 German cockroach
Blatta germanica
SIZE AND DESCRIPTION
10–13 mm long. Pale reddish-brown. The long wings extend beyond the tip of the abdomen. There are two dark, longitudinal marks on the pronotum (the shield covering the head). Can fly, but rarely does so.
HABITAT Buildings, refuse-tips in summer. Introduced from North Africa.
FOOD/HABITS Scavenges for food on the ground.

2 Common earwig
Forficula auricularia
SIZE AND DESCRIPTION
Body length: 10–15 mm; pincers measure 4–9 mm in the male and 4–5 mm in the female.
HABITAT Abundant throughout Europe, in a wide range of habitats. Very common in gardens.
FOOD/HABITS Mainly vegetarian. White earwigs found in the garden are in the process of moulting. Displays parental care for its young when disturbed.

3 Small earwig
Labia minor
SIZE AND DESCRIPTION
Body length: 5 mm; pincers up to 2.5 mm long. Dull, darkish-brown body and a blackish head. Hindwings extend beyond forewings when at rest.
HABITAT Found anywhere there is decaying vegetation. Common around compost heaps.
FOOD/HABITS Flies well, mostly at dusk. Breeds in manure heaps. Displays parental care for young.

4 Common cockroach
Blatta orientalis
SIZE AND DESCRIPTION
18–30 mm long. Male's leathery wings extend to the last three segments of the abdomen; female's wings barely cover the thorax.
HABITAT Warm indoor places, such as kitchens. Rubbish tips in summer. Survives outdoors in mild parts of Europe. Originates from Asia and Africa.
FOOD/HABITS Scavenges on the ground for food scraps and decaying matter.

5 Praying mantis
Mantis religiosa
SIZE AND DESCRIPTION
20 mm long. Green body, sometimes brown. Males are particularly slender.
HABITAT Rough grassland, scrub and maquis in Europe, as far south as southern France.
FOOD/HABITS Preys on other insects. Adopts a threat display, raising the neck and front legs in a "praying" posture. Female eats the male after or during copulation.

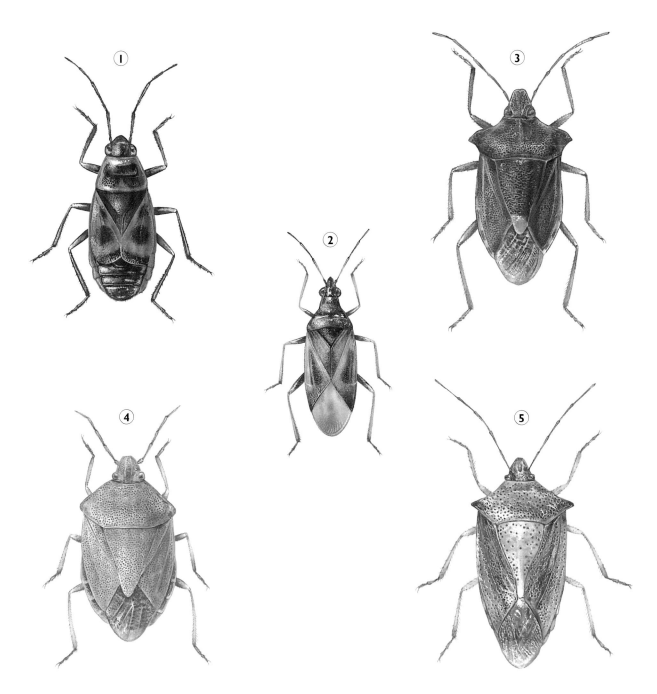

1 Fire bug
Pyrrhocoris apterus
SIZE AND DESCRIPTION
10 mm long. Red and black, with black spots on the red forewings.
HABITAT Found in open country in central and southern Europe.
FOOD/HABITS The fire bug is omnivorous, feeding on fallen seeds and preying upon other insects. Adults hibernate.

2 Common flower bug
Anthocoris nemorum
SIZE AND DESCRIPTION
3–4 mm. Shiny and generally brownish, with a black spot on the greyish forewings. Head is black.
HABITAT Found on almost any type of tree, shrub or herbaceous plant. Occurs over most of Europe.
FOOD/HABITS A predator of aphids, red spider mites and other insects. Adults hibernate under loose bark and in clumps of grass.

3 Forest bug
Pentatoma rufipes
SIZE AND DESCRIPTION
12–15 mm long. Dark brown, with a thorax that protrudes like a yoke. Reddish legs.
HABITAT Found on trees (particularly cherry trees) in orchards and shrubberies.
FOOD/HABITS Adults are seen between June and October. An omnivorous bug that sucks juice from buds, leaves and fruits, and attacks other insects.

4 Green shield bug
Palomena prasina
SIZE AND DESCRIPTION
10–15 mm long. Bright green in spring and summer, bronze-coloured in autumn. Wing tips are dark brown.
HABITAT Woodland edges and glades, hedgerows and gardens with shrubs and herbaceous borders over much of Europe.
FOOD/HABITS Eats leaves of trees, shrubs and herbaceous plants. Hibernates in leaf litter.

5 Hawthorn shield bug
Acanthosoma haemorrhoidale
SIZE AND DESCRIPTION
15 mm long. Body is shield-shaped, with a reddish-brown band along the rear of the thorax.
HABITAT Woodland edges, hedgerows and gardens with hedges and shrubs. Widespread in Europe, but absent from Scotland.
FOOD/HABITS Eats leaves of hawthorn and fruit trees. Basks on walls in autumn before hibernating.

33

Bugs

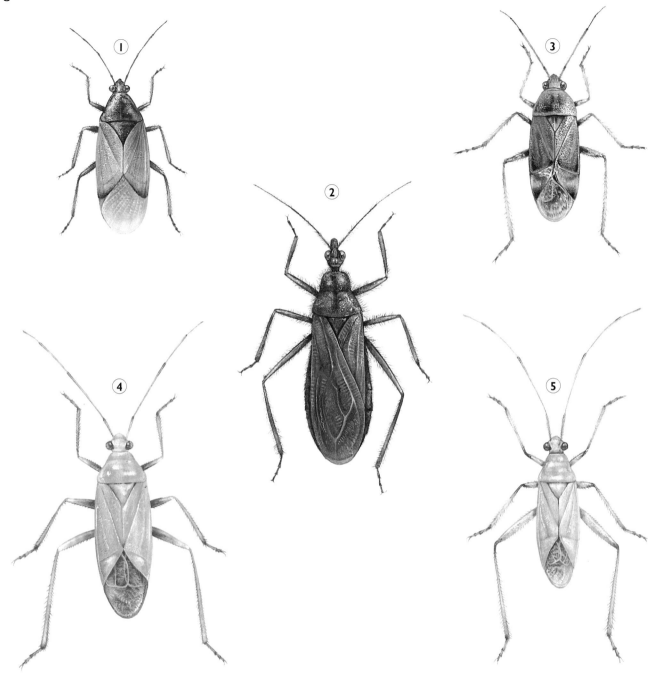

1 Hot-bed bug
Xylocoris galactinus
SIZE AND DESCRIPTION
6 mm long. Brownish wings with translucent tips. Through a lens, it is possible to see that the bug's antennae are hairy. HABITAT Compost-heaps, hot-beds and other warm soil. Also found in birds' nests and stables. FOOD/HABITS Feeds on other insects, and will suck the blood of birds.

2 Fly bug
Reduvius personatus
SIZE AND DESCRIPTION
17 mm long. Dark brown to black. Bristly legs and abdomen. Wings extend to the tip of the abdomen. HABITAT Places with plenty of crevices; most common around human habitation. Found in much of Europe, including southern England. FOOD/HABITS Nocturnal feeder on small insects, including bed bugs. Makes a "squeak" by rubbing the tip of its beak against its thorax. Will stab people with its beak if handled.

3 Tarnished plant bug
Lygus rugulipennis
SIZE AND DESCRIPTION
4–6 mm long. Variable colour, from yellow to red or brown. Wing-tips are membranous. Rather bristly legs. HABITAT Gardens and other places with plenty of vegetation. Occurs over most of Europe. FOOD/HABITS Feeds on potatoes and other crops, as well as flowers and stinging nettles. Causes white spots on leaves. Winters in leaf litter. Most common in late summer.

4 Common green capsid
Lygocoris pabulinus
SIZE AND DESCRIPTION
5–7 mm long. Green, with greyish wing-tips. Bristly legs. The potato capsid is similar, but has two black spots behind its head. HABITAT Across most of Europe, wherever there is plentiful vegetation. FOOD/HABITS Flies from May to October. Eggs laid on woody plants hatch in spring. Eats herbaceous plants, potatoes and soft fruits such as raspberries and gooseberries.

5 Black-kneed capsid
Blepharidopterus angulatus
SIZE AND DESCRIPTION
15 mm long. Green, with a narrower body than that of the common green capsid. Legs have black patches on the "knees". HABITAT Orchard trees, particularly apples and limes. FOOD/HABITS This predatory insect is beneficial to orchard owners, as it feeds on red spider mites, which cause damage to fruit trees.

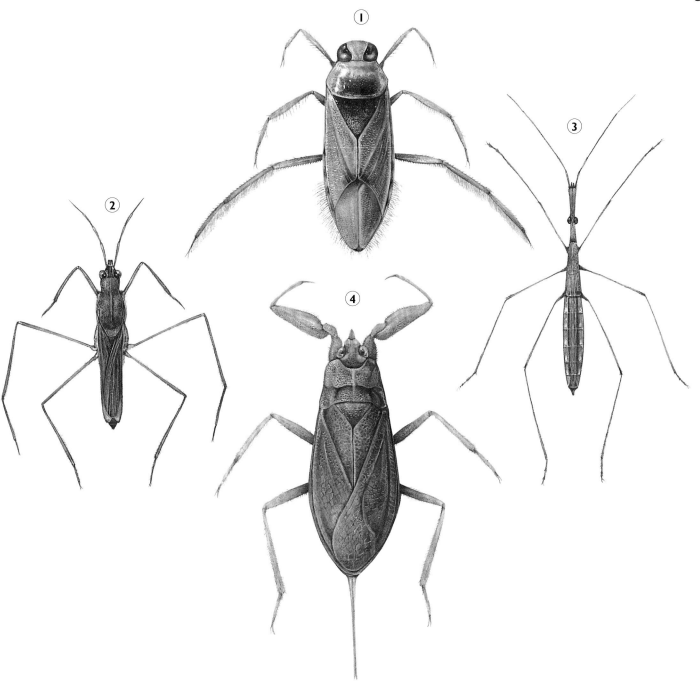

1 Common backswimmer

Notonecta glauca

SIZE AND DESCRIPTION

16 mm long. Has long, bristly hindlegs. Swims on its back, which is keeled, clutching a large air-bubble to its "underside". There are several species of water boatman.

HABITAT Swims in still water, and will fly in warm weather.

FOOD/HABITS A hunter of tadpoles, small fish and other insects. Active all year round.

2 Common pond skater

Gerris lacustris

SIZE AND DESCRIPTION

10 mm long. Has a broader body than the water measurer and a considerably shorter head, which has largish eyes. Usually fully-winged. Several similar species.

HABITAT Lives on the surface of slow-moving water.

FOOD/HABITS Flies away from water to hibernate. When swimming, it moves across the water's surface with a rowing action of the middle legs. The trailing hindlegs act as rudders, while the front legs catch insects that fall into the water.

3 Water measurer

Hydrometra stagnorum

SIZE AND DESCRIPTION

11 mm long, with a narrow body, greatly elongated head and long legs. Usually without wings.

HABITAT Found on the surface of still or slow-flowing water across much of Europe.

FOOD/HABITS Feeds on water fleas, insect larvae and other small animals, which it spears from the surface with its beak. Moves slowly across the surface.

4 Water scorpion

Nepa cinerea

SIZE AND DESCRIPTION

Body is 20 mm long, with a tail measuring 8 mm. The brown, flattened body is equipped with strong front legs. Fully-winged, but rarely flies.

HABITAT Lives in shallow water and at pond margins.

FOOD/HABITS Breathes through the hollow, snorkel-like "tail", which draws in air as it protrudes above the surface. Active throughout the year, the water scorpion feeds on invertebrates and small fish, which are caught with the powerful front legs.

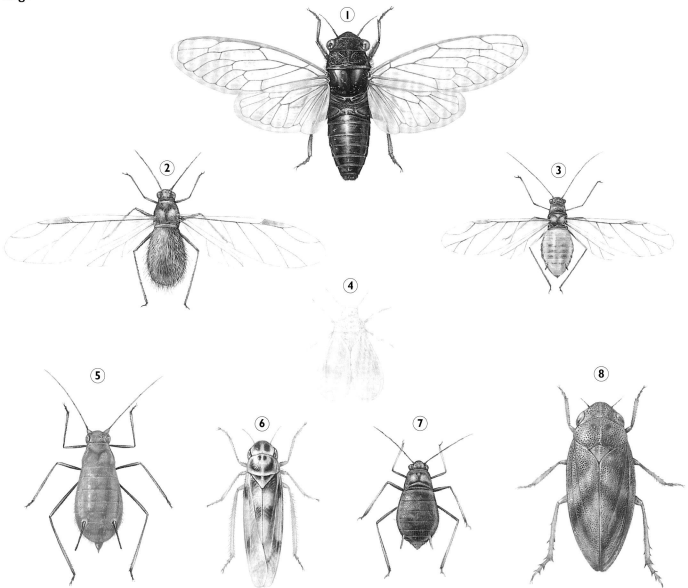

1 Cicada
Cicada montana
SIZE AND DESCRIPTION 18 mm long.
Long transparent and shiny wings.
Dark pronotum. Three spines on
front femur.
HABITAT Woodland clearings
and scrubby areas in central and
southern Europe.
FOOD/HABITS Flies from May
to August. Feeds on shrubs and
herbaceous plants.

5 Rose aphid
Macrosiphum rosae
SIZE AND DESCRIPTION 1–2 mm
long. This "greenfly" can be
either green or pink. Long black
cornicles on abdomen not found
on other aphids.
HABITAT Woodland edges, hedges
and gardens across Europe.
FOOD/HABITS Feeds on roses in
spring and scabious or teasel
in summer.

2 Woolly aphid
Eriosoma lanigerum
SIZE AND DESCRIPTION
1–2 mm long. Purplish-brown, with
or without wings, and covered
with strands of whitish, fluffy wax.
HABITAT Orchards and gardens
across Europe. Accidentally
introduced from America.
FOOD/HABITS Sucks the sap of
fruit trees. Most young are born
live by parthogenesis.

6 Potato leafhopper
Eupteryx aurata
SIZE AND DESCRIPTION
4 mm long. Black-and-yellow
pattern, often orange-tinged.
Wings reach past abdomen's tip.
HABITAT Wasteland, gardens and
hedgerows across Europe, except
in the far north.
FOOD/HABITS Sucks sap from
herbaceous plants. Adults are
seen from May to December.

3 Cabbage aphid
Aleyrodes proletella
SIZE AND DESCRIPTION
2 mm long. Winged individuals
are dark green and black.
Wingless individuals have a
mealy-white covering.
HABITAT Wasteland and farmland
throughout Europe. Abundant in
spring and summer.
FOOD/HABITS Feeds on brassica
plants in spring and early summer.

7 Black bean aphid
Aphis fabae
SIZE AND DESCRIPTION
2 mm. Black or olive in colour.
May be wingless.
HABITAT All over Europe where
there are suitable food plants.
FOOD/HABITS Feeds on young
shoots of dock, beans, spinach,
beet, nasturtium and other plants.
Eggs are laid on shrubs such as
spindle and Philadelphius.

4 Cabbage whitefly
Ateryrodes protella
SIZE AND DESCRIPTION
2–3 mm long. Waxy-white wings.
HABITAT Fields and gardens
all over Europe. Found on the
underside of cabbage leaves.
FOOD/HABITS Sucks sap from
cabbage leaves. Active all year.

8 Common froghopper
Philaenus spumarius
SIZE AND DESCRIPTION
6 mm long. Variable brown
pattern. Wings held together like
a tent. Young coat themselves in a
white broth called "cuckoo-spit".
HABITAT Woody and herbaceous
plants across Europe, except in
the far north.
FOOD/HABITS Flies from June to
September. Feeds on plant sap.

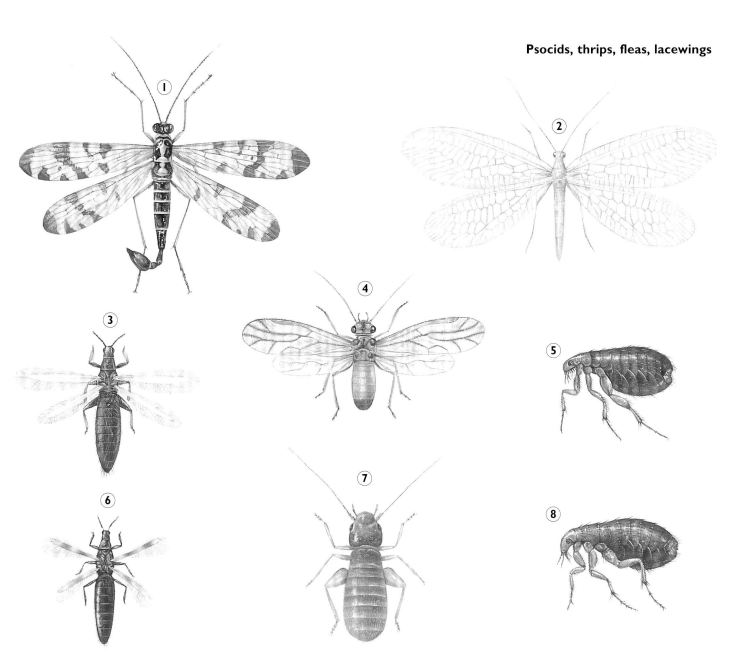

1 Scorpion fly
Panorpa communis
SIZE AND DESCRIPTION 15 mm long; wingspan of 30 mm. Head has a "beak". Scorpion-like tip to male's abdomen.
HABITAT Woods, hedgerows and shaded gardens throughout Europe, except in the far north.
FOOD/HABITS Flies (weakly) from May to August. Carnivorous. Larvae scavenge in soil.

2 Green lacewing
Chrysopa pallens
SIZE AND DESCRIPTION 15–20 mm long; wingspan of 30–40 mm. Bright green body, golden eyes and green veins on transparent wings. Several continental species; two similar species in British Isles.
HABITAT Woods, hedgerows, gardens and well-vegetated areas. Most of Europe, but not Scotland and northern Scandinavia.
FOOD/HABITS Mainly nocturnal. Adults and young prey on aphids, Flies May to August.

3 Pea thrips
Kakothrips pisivorus
SIZE AND DESCRIPTION 2.5 mm long. Black body, with white feathery wings.
HABITAT Well-vegetated open areas, gardens and allotments.
FOOD/HABITS Scrapes pea-pods to reach their sap, leaving silvery patches on the pods where it has been feeding.

4 Bark louse
Caecilius flavidus
SIZE AND DESCRIPTION 1–2 mm long. Body is yellow, and the wings are bristly.
HABITAT Broad-leaved trees. Very common throughout Europe.
FOOD/HABITS Algae and pollen on the foliage of trees. Males are unknown. Females reproduce by parthenogenesis.

5 Cat flea
Ctenocephalides felis
SIZE AND DESCRIPTION 3 mm long.
HABITAT Found mostly in houses with cats and dogs. Breeds rapidly in warm weather.
FOOD/HABITS Blood-suckers. Will bite humans and other mammals.

6 Thunder-fly
Aeolothrips intermedius
SIZE AND DESCRIPTION 2.5 mm long. Blackish body, with brown-barred wings.
HABITAT Found in well-vegetated open areas.
FOOD/HABITS Eats sap of yellow crucifers and compositae.

7 Booklouse
Liposcelis terricolis
SIZE AND DESCRIPTION 1–2 mm. Brown and flat-bodied, with a large head. Several similar species.
HABITAT Occurs indoors.
FOOD/HABITS Feeds on books, paper and stored food.

8 Human flea
Pulex irritans
SIZE AND DESCRIPTION 3.5 mm long. Very similar to cat flea.
HABITAT Buildings, human beings, badgers and foxes.
FOOD/HABITS Feeds on a variety of mammals. Transmit several diseases, including "black death".

1 Orange tip
Anthocaris cardamines
SIZE AND DESCRIPTION
Forewing is 20–25 mm.
Male has orange wing-tips
and green-blotches on
underside of hindwing.
Female has greyish patches
on forewing, and mottled
underwings.
HABITAT Hedgerows,
gardens, damp meadows
and woodland margins.
All Europe, except south-
west or southern Spain or
northern Scandinavia.
FOOD/HABITS Flies April
to June. Larvae eat garlic
mustard, lady's smock,
but also sweet rocket
and honesty in gardens.
Overwinters as a pupa.

2 Small white
Artogeia rapae
SIZE AND DESCRIPTION
Forewing is 15–30 mm.
Upperside is white, with
one black or grey spot on
male's forewing and two
on female's. Black or grey
forewing patches extend
further along the leading
edge than down the side
of the wing. Two spots
on underside of forewing
in both sexes. Underside
of hindwings is yellowish.
The larva is green, with a
yellow stripe running along
its side.
HABITAT Gardens, hedges
and flowery places across
Europe. Abundant to the
point of being a pest.
FOOD/HABITS Flies March
to October. Two to four
broods. Eggs are laid on
leaves. Larvae feed on
brassicae and nasturtiums.

3 Green-veined white
Artogeia napi
SIZE AND DESCRIPTION
Forewing is 18–30 mm.
Black spots and patches
on the forewing are less
distinct than in the small
or large white. Grey lines
along veins on underside
of hindwings. Larva is
similar to that of small
white, without the yellow.
HABITAT Gardens,
hedges, woodland
margins and other
flowery places throughout
Europe.
FOOD/HABITS Flies March
to November. Larvae eat
crucifers, such as garlic
mustard, lady's smock
and watercress.

4 Brimstone
Gonepteryx rhamni
SIZE AND DESCRIPTION
Forewing is 25–30 mm.
Male's wings are sulphur
yellow on top, but paler
beneath. Female is white,
with a pale green tinge,
but she lacks the large
white's black markings.
Larva is green, with white
stripes along the side.
HABITAT Open woodland,
gardens and flowery places.
All Europe, but not most
of Scotland and northern
Scandinavia.
FOOD/HABITS Flies
February to September.
Larvae eat buckthorn and
alder buckthorn. Adults
overwinter in holly or ivy.

5 Large white
Pieris brassicae
SIZE AND DESCRIPTION
Forewing is 25–35 mm.
Black tips extend halfway
down the forewing's edge.
Upperside of forewing has
two black spots in female,
one in male. Underside of
forewing has two spots
in both sexes. Caterpillar
is green, with black spots
and yellow stripes.
HABITAT Gardens and
other flowery places.
FOOD/HABITS Flies April
to October. Eggs are
laid on the underside of
leaves. The larvae feed
on brassicae and
nasturtiums.

1 Small copper
Lycaena phlaeas
SIZE AND DESCRIPTION
Forewing is 10–17 mm.
Bright forewing is like
shiny copper, with dark
flecks and brown edges.
Caterpillar is small and
green.
HABITAT Gardens,
flowery wasteland and
heathland across Europe.
FOOD/HABITS Flies
February to November.
Two or three broods,
with adults from the third
brood being rather small.
Food plants for larvae are
common sorrel, sheep's
sorrel and docks.

2 Small heath
Coenonympha pamphilus
SIZE AND DESCRIPTION
Forewing is 14–16 mm.
Orange-brown, with very
small dark "eyes". Green
larva has a white stripe
along the side.
HABITAT Grassy places
up to 2,000 m. Across
Europe, except northern
most Scandinavia.
FOOD/HABITS Flies April
to October. Pupa feeds
on grasses. One to three
broods. Winters as
a larva.

3 Small skipper
Thymelicus flavus
SIZE AND DESCRIPTION
Forewing is 13–15 mm.
Bright orange wings. The
body is stout and rather
moth-like. Tends to hold
its wings flat when
at rest.
HABITAT Grassy places.
England, Wales and
mainland Europe south
from Denmark.
FOOD/HABITS Flies May
to August. Swift, darting
flight. The small green
larvae feed briefly on
grasses, but go into
hibernation almost
immediately after
hatching.

4 Holly blue
Celastrina argiolus
SIZE AND DESCRIPTION
Forewing is 12–18 mm.
Upperside of male is violet
blue. Female is paler blue,
edged with a broad dark
band. The dark band is
broader in the second
brood. Underside of the
wings is pale blue-grey. The
caterpillar is small, green
and slug-like. This is the
blue most likely to be seen
in gardens.
HABITAT Woodland
margins, hedgerows, parks
and gardens. Found across
Europe, except in Scotland
and northern Scandinavia.
FOOD/HABITS Flies April
to September. First brood
feeds on flowers and
developing fruit of holly;
second brood feeds on ivy.
Adults drink honeydew,
sap and juices of carrion.
Winters as a pupa.

5 Common blue
Polyommatus icarus
SIZE AND DESCRIPTION
Forewing is 14–18 mm.
Upperside of male's
wings are violet blue; the
upperside of the female's
wings are dark brown.
Small greenish larva.
HABITAT Flowery
grasslands, roadsides,
sand dunes and
wasteland. Occurs
throughout the whole
of Europe.
FOOD/HABITS Flies
April to October. Two
or three broods. Food
consists of leguminous
plants, particularly
horseshoe vetch. Winters
as a small larva.

Butterflies

1 Comma
Polygonia c-album
SIZE AND DESCRIPTION
Forewing is 23 mm. Wings have jagged edges. Orange upperside, with black and buff markings. Underside of hindwing has a white comma-shaped mark. Caterpillar is black and sparsely bristled. Its rear end becomes white, making it look like a bird dropping.
HABITAT Woodland margins, gardens, hedges and other flowery places. Common across Europe, but absent from Ireland, northern Britain, and northern Scandinavia.
FOOD/HABITS Flies March to September. There are two broods. Second brood is darker. Overwinters, with adults hanging from leaves. Larvae feed on stinging nettles, hops and elms.

2 Painted lady
Cynthia cardui
SIZE AND DESCRIPTION
Forewing is 20–25 mm. Upperside is orange, with a black forewing tip patched with white. Underside is pale, with three blue underwing spots. Black caterpillar has tufts of hairs and a yellow-and-red stripe down the side.
HABITAT Flowery places, including roadsides and gardens. Across Europe, but is a migrant from North Africa. Does not survive European winter.
FOOD/HABITS Flies April to November, arriving in Britain in late spring/early summer. Two broods in Europe, but produces broods throughout the year in North Africa. Eats thistles and sometimes stinging nettles.

3 Small tortoiseshell
Aglais urticae
SIZE AND DESCRIPTION
Forewing is 25 mm. Upperside is bright orange and black, with a row of blue spots on the trailing edge of the hindwings. Caterpillar is bristly and black.
HABITAT All kinds of flowery places. Common across the whole of Europe.
FOOD/HABITS Flies March to October. Adults overwinter, often in buildings. Larvae feed on nettles, elms and hops.

4 Red admiral
Vanessa atalanta
SIZE AND DESCRIPTION
Forewing is 30 mm. Upperside is a velvety dark brown, with bright orange bars on each wing. Tips of forewings are black with white markings. Underside of hindwing is pale brown, while underside of forewing shows orange, blue and white markings. Dark caterpillar has bristles and a pale yellow stripe along the side.
HABITAT Flowery places across Europe. Absent from northern Scandinavia. Resident in southern Europe, moves north in spring.
FOOD/HABITS Flies May to October. There are two broods. Larvae feed on nettles. Adults feed on rotting fruit in autumn.

5 Peacock
Inachis io
SIZE AND DESCRIPTION
Forewing is 30 mm. Wings have four large, peacock-like "eyes". Upperside is orange, while underside is very dark brown. Caterpillar is black and bristly.
HABITAT Flowery places, including gardens. Across Europe, and as far north as southern Scandinavia.
FOOD/HABITS Flies March to May, and July to September. Larvae feed on nettles. Adults often overwinter in buildings.

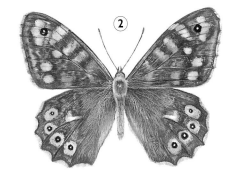

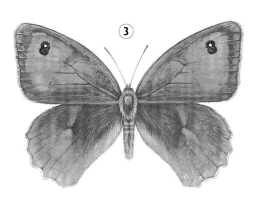

1 Gatekeeper
Pyronia tithonus
SIZE AND DESCRIPTION Forewing is
17–25 mm. Usually smaller than
the meadow brown, with orange
patches on the wings. "Eyes" are
black with two highlights. Green
or brown larva.
HABITAT Hedgerows and
woodland margins. Southern
Britain and Ireland, and south
across the rest of Europe.
FOOD/HABITS Flies July to
September. Larval food plants
are fine-leaved grasses. Adults
are fond of bramble blossom
and marjoram.

2 Speckled wood
Pararge aegeria
SIZE AND DESCRIPTION Forewing
is 19–22 mm. Yellow- or orange-
and-brown wings. Orange-spotted
form, *P. aegeria*, south-west
Europe and Italy; cream-spotted
form, *P. a. tircis*, elsewhere.
Green caterpillar.
HABITAT Woodland clearings,
gardens and paths across Europe
from southern Scandinavia.
FOOD/HABITS Flies March to
October. Feeds on grasses.
Single-brooded in north.
Overwinters in both larval and
pupal forms.

3 Meadow brown
Maniola jurtina
SIZE AND DESCRIPTION Forewing
is 20–26 mm. Brown and orange.
Upper wing has a single black eye
with a white highlight. Females are
larger than males. The green larva
has a white stripe along
the side.
HABITAT Grassland; also
woodland in southern Europe.
Very common across Europe
southwards from southern
Scandinavia up to 2,000 m.
FOOD/HABITS Flies May to
September. Larvae feed on grasses.
Winters as a larva.

4 Wall brown
Lasiommata megera
SIZE AND DESCRIPTION Forewing
is 17–25 mm. Brown-and-orange
patterned, with "eyes" on the
forewings. Underside of
hindwing is pale silvery-brown.
Green caterpillar.
HABITAT Rough grassy places and
gardens. British Isles (not nothern
Scotland) and across from
southern Europe.
FOOD/HABITS Flies March to
October. Adults sunbathe on
walls and fences. Food plants are
grasses. Two or three broods.
Overwinters as a larva.

Moths

1 Codlin moth
Cydia pomonella
SIZE AND DESCRIPTION
Forewing is about 10
mm. Grey forewing has
black and yellowish marks
towards the tips. White
larva has a brown head,
becoming pinkish as it
grows larger.
HABITAT Orchards, parks,
gardens and hedges with
apple trees. All Europe,
except the far north.
FOOD/HABITS Flies May
to October. Two broods.
Larvae bore into apples
(and pears) to eat both
the flesh and developing
seeds. Pupates under
loose bark in a cocoon,
from which it emerges
during spring.

2 Tapestry moth
Trichophaga tapetzella
SIZE AND DESCRIPTION
Forewing is 8–10 mm.
Greyish-white wings have
brown patches towards
the thorax.
HABITAT Found in stables
and buildings with high
humidity.
FOOD/HABITS Flies in June
and July. The tapestry is
a clothes moth whose
larvae feed on animal
fibres such as wool.
Found in horsehair stuffing
and owl pellets.

3 Leaf-mining moth
Stigmella aurella
SIZE AND DESCRIPTION
Forewing is 3 mm.
The forewings have
yellow bars and purplish
wingtips. Pale, feathery
underwings. The tiny,
leaf-mining larva creates
pale, squiggly lines on
bramble leaves.
HABITAT Woodland,
hedges and gardens over
most of Europe, except
the far north.
FOOD/HABITS Flies May
to September. The food
plant is bramble. Larvae
overwinter in the
leaf-mines, but leave
before pupating.

**4 Common clothes
moth**
Tineola bisselliella
SIZE AND DESCRIPTION
Forewing is 4–6 mm.
Goldish forewings and
silvery hindwings. The
white larva has a pale
brown head.
HABITAT Rarely seen
outdoors. The most
common and destructive
clothes moth.
FOOD/HABITS Adults
are found throughout
the year. Rarely flies,
preferring to scuttle for
cover. Larvae eat animal
fibres and also build
shelters out of them.

5 *Lampronia rubiella*
SIZE AND DESCRIPTION
Forewing is 5 mm. Two
yellow or cream bars on
the dark brown wings.
HABITAT Gardens with
raspberries. Found across
northern and central
Europe.
FOOD/HABITS Flies in May
and June. Larvae feed
in the central stalks of
raspberry fruit in summer.
They overwinter in the
soil and then complete
their growth
in buds during spring.

1 Green oak tortrix moth
Tortrix viridana
SIZE AND DESCRIPTION
Forewing is about 10 mm. Pale green forewings and pale grey hindwings. Green larva measures about 12 mm long.
HABITAT Woods, parks and gardens with oaks.
FOOD/HABITS Flies May to August at night, but lives for only one week. The larvae feed on the buds and rolled leaves of oak trees. Will hang on a thread from trees.

2 Gold fringe
Hypsopygia costalis
SIZE AND DESCRIPTION
Forewing is 8 mm. The dark brown forewings have two gold marks and and a golden yellow fringe. The hindwings are purplish with a gold fringe. Larva is whitish with a brown head.
HABITAT Hedges around grassy places in southern Britain and south/central Europe.
FOOD/HABITS Flies July to October. Larvae feed on dead grasses and thatch.

3 White plume
Pterophorus pentadactyla
SIZE AND DESCRIPTION
Forewing is 12–15 mm. White, with each forewing being split into two feathery sections and each hindwing into three feathery sections. The bright green larva has tufts of silvery hair.
HABITAT Hedgerows, waste ground and gardens across Europe.
FOOD/HABITS Flies May to August at night. Larvae feed on bindweeds, curled up in a leaf. Overwinters as a small caterpillar.

4 Small magpie
Eurrhypara hortulata
SIZE AND DESCRIPTION
Forewing is about 15 mm. Silky white, with dark grey markings and a yellowish-gold thorax with black spots. Green caterpillar.
HABITAT Hedgerows, woodland margins and waste ground with nettles. All Europe, except the far north.
FOOD/HABITS Flies June to August. Larvae feed on stinging nettles and related plants. Winters as a larva in a spun cocoon among plant debris.

5 Common swift
Hepialus lupulinus
SIZE AND DESCRIPTION
Forewing is 12–18 mm. Brown wings with white marks. Very short antennae. Wings are held tightly to the body when at rest. White or creamy larva is about 35 mm long with a brown head.
HABITAT Arable land, gardens, parks and grassland over most of Europe, except Iberia.
FOOD/HABITS Flies May to August at dusk. The caterpillar lives in soil, eating the roots of grasses and other herbaceous plants. Overwinters as a caterpillar.

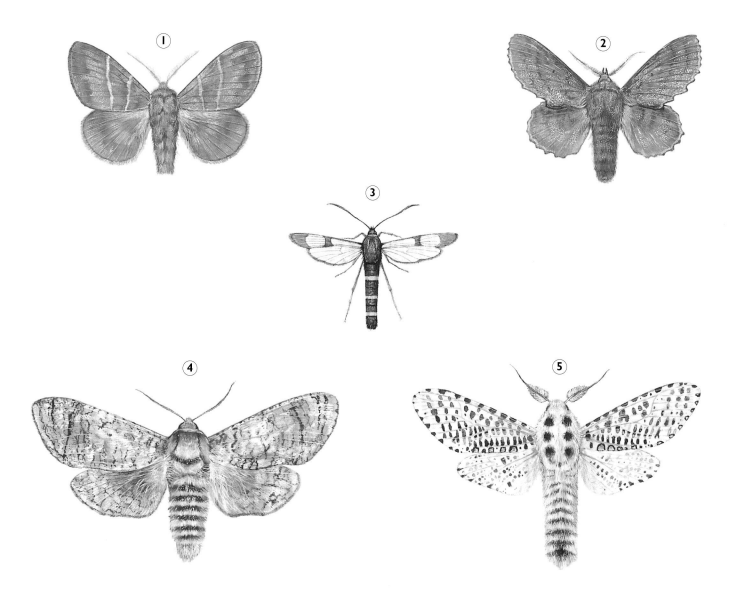

1 Fox moth
Macrothylacia rubi
SIZE AND DESCRIPTION
Forewing is 20–30 mm.
Males are fox-coloured,
with two narrow pale
strips on the forewings;
females are paler. The
larva is velvety and very
dark brown, with
orange bands.
HABITAT Heathland,
moorland, open
countryside and
woodland margins.
FOOD/HABITS Flies May
to July. The male flies in
sunshine and at night,
the female is a purely
nocturnal flier. Larvae
feed on bramble, heather,
bilberry and creeping
willow. Overwinters as a
full-grown larva.

2 Lappet moth
Gastropacha quercifolia
SIZE AND DESCRIPTION
Forewing is about 30 mm.
Females are much larger
than males. Varies across
its range, from purple in
the north to pale brown
in south. At rest, the wings
resemble dead leaves.
Larva is dark brownish-
grey, with two bluish
bands near its head.
HABITAT Open woodland,
hedges, orchards and
gardens.
FOOD/HABITS Flies May
to August at night. Single-
brooded. Larvae feed on
blackthorn, buckthorn,
apples and sallows.
Overwinters as a larva.

3 Currant clearwing
Synanthedon tipuliformis
SIZE AND DESCRIPTION
Forewing is about 8
mm. Wings are largely
transparent. Black
abdomen has four yellow
rings in the male and three
in the female. The dingy
white larva has a brown
head and yellow spots. The
currant clearwing usually
rests with its wings apart.
HABITAT Woods, gardens
and open country with
suitable foodplants across
Europe.
FOOD/HABITS Flies May
to July. Single-brooded.
Larvae feed inside the
stems of blackcurrant,
whitecurrant and
gooseberry. Winters as
a larva.

4 Goat moth
Cossus cossus
SIZE AND DESCRIPTION
Forewing is about 30
mm. Greyish-brown
wings with fine dark
patterning on forewings.
The abdomen is distinctly
ringed. Large and solid-
looking. The purplish-red
larva emits a goat-like
smell.
HABITAT Broad-leaved
woodland from Ireland
and England south across
Europe.
FOOD/HABITS Flies June
to August. The larvae
feed on the solid wood
of broad-leaved trees for
about three years before
pupating in the ground.

5 Leopard moth
Zeuzera pyrina
SIZE AND DESCRIPTION
Forewing is 20–35 mm.
White, with finely spotted
wings and six black marks
on its furry thorax. The
abdomen is ringed with
greyish black. Females
are much larger than
males. The creamy larva
has black spots and a dark
head.
HABITAT Woods, parks,
orchards and gardens.
Found across central and
southern Europe from
England.
FOOD/HABITS Flies June
to August at night. Single-
brooded. Larvae tunnel
into broad-leaved trees
and shrubs.

1 Peach blossom
Thyatira batis
SIZE AND DESCRIPTION
Forewing is about 15 mm. Forewings are brown with pink blotches. The larva is dark brown, with slanting white lines and bumps on its back.
HABITAT Woodland and woodland edges in northern and central Europe, including the British Isles.
FOOD/HABITS Flies May to August at night. Single-brooded. Larvae feed on bramble, raspberry and blackberry. Overwinters as a pupa.

2 Lime-speck pug
Eupithecia centaurearia
SIZE AND DESCRIPTION
Forewing is about 12 mm. Pale grey, with dark marks on the forewings. The green or yellow larva often has red spots.
HABITAT Rough areas and gardens.
FOOD/HABITS Flies May to October from dusk. Rests with its wings outstretched on lichens on walls and tree-trunks. May be double-brooded. Larvae feed on a range of herbaceous plants, such as yarrow and ragwort. Overwinters as a pupa.

3 Garden carpet
Xanthorhoe fluctuata
SIZE AND DESCRIPTION
Forewing is about 14 mm. Greyish-white wings, with markings in various shades of grey. Pattern varies, but there is always a dark triangle where forewings join thorax. Larva is a twig-like looper whose colour ranges from grey-green to dark brown.
HABITAT Common in cultivated areas.
FOOD/HABITS Flies April to October from dusk. Rests on walls and fences during the day. Two or three broods. Larvae feed on perennial wall-rocket, garlic mustard and other crucifers. Winters as a pupa.

4 Small emerald
Hemistola chrysoprasaria
SIZE AND DESCRIPTION
Forewing is 18 mm. Pale grey-green, with two fine white lines on the forewing and a single line on the hindwing. The larva is pale green, with white dots and a brown head.
HABITAT Downland, hedges and woodland edges, usually on chalk or limestone.
FOOD/HABITS Flies May to August at night. Single-brooded. Larvae feed on traveller's joy.

5 Lackey moth
Malacosoma neustria
SIZE AND DESCRIPTION
Forewing is 13–20 mm. Comes in a range of browns. Similar to the fox moth, but wing bands curve inwards. Long, tufted, grey-blue larva has white, orange, black and yellow stripes along its body.
HABITAT Many habitats over most of Europe, except Scotland and northern Scandinavia.
FOOD/HABITS Flies June to August at night. Single-brooded. Larvae live in colonies in cocoons, feeding on the leaves of hawthorn, blackthorn, plums and sallows. Winters as an egg.

1 Peppered moth
Biston betularia
SIZE AND DESCRIPTION Forewing is 20–30 mm. Variable. Normal form is white, peppered with fine dark marks, or sooty black. The green or brown looper caterpillar is up to 60 mm long.
HABITAT Woods, gardens, scrub and parks across Europe, except the far north.
FOOD/HABITS Flies May to August, coming to lighted windows. Larvae feed on a range of trees and shrubs, including sallow, hawthorn, golden rod and raspberry.

4 August thorn
Ennomos quercinaria
SIZE AND DESCRIPTION Forewing is about 17 mm. Pale yellowish-tan, with two narrow brown stripes on each forewing. Abdomen is fluffy. The greyish-brown looper caterpillar has nodules that make it look like a twig.
HABITAT Woodland, parks and gardens. Locally common south from Scotland.
FOOD/HABITS Flies August and September at night. Larvae feed on oaks and other trees.

2 Winter moth
Operophtera brumata
SIZE AND DESCRIPTION Forewing is about 15 mm. Males have greyish-brown, faintly patterned wings; females have stunted, relict wings. The green looper caterpillar is about 20 mm long.
HABITAT Abundant wherever there are trees and shrubs.
FOOD/HABITS Flies October to February. Nocturnal and attracted to lighted windows. Females can be seen on windowsills and tree-trunks. Larvae feed on deciduous trees. A serious pest of hard fruits, especially apples.

5 Brimstone
Opisthograptis luteolata
SIZE AND DESCRIPTION Forewing is 15–20 mm. Sulphur yellow, with brown flecks on leading edge of forewing. Brown looper caterpillar is about 30 mm long, often tinged grey or green, with a prominent nodule on the middle of its back.
HABITAT Woods, hedges and gardens.
FOOD/HABITS Flies April to October at night. Caterpillar feeds on hawthorn, blackthorn and other shrubs.

3 Magpie moth
Abraxas grossulariata
SIZE AND DESCRIPTION Forewing is about 20 mm. Variable black-and-white pattern, with a yellowish-orange line across the middle of the forewing and near the head. The larva, about 30 mm long, is pale green with black spots and a rusty line along its sides.
HABITAT Woods, gardens and hedges. FOOD/HABITS Flies June to August. Larvae feed on blackthorn, currants, hawthorn and many other shrubs. Overwinters as a small caterpillar and pupates in May or June.

6 Swallowtailed moth
Ourapteryx sambucaria
SIZE AND DESCRIPTION Forewing is 25–30 mm. Wings are bright lemon, rapidly fading to pale cream or white. Caterpillar is a brown looper and up to 50 mm long.
HABITAT Forest edges, woods, gardens, scrub and parks across Europe, except the far north.
FOOD/HABITS Flies June to August at night, often coming to lighted windows. Larvae feed on blackthorn, hawthorn, ivy and numerous other trees and bushes.

1 Hummingbird hawkmoth
Macroglossum stellatarum
SIZE AND DESCRIPTION Forewing is about 25 mm. Mousey-grey forewings and a hairy thorax. Hindwings are golden orange. The caterpillar, about 50 mm long, is green with yellow, white and green horizontal stripes.
HABITAT Parks, gardens and flowery banks. Found in southern Europe, moving northwards in summer, reaching Britain in varying numbers.
FOOD/HABITS Day-flying throughout the year. Usually seen in summer in Britain. Hovers in front of flowers, drinking nectar through its long proboscis. Caterpillars feed on bedstraws.

2 Eyed hawkmoth
Smerinthus ocellatus
SIZE AND DESCRIPTION Forewing is up to 45 mm. There is a wavy trailing edge to the forewing; the hindwing is pink, with a large blue "eye". The bright green caterpillar has seven diagonal yellow stripes on each side of its body and a greenish-blue horn at the rear.
HABITAT Open woodland, parks and gardens across Europe, but not Scotland or northern Scandinavia.
FOOD/HABITS Flies May to July. Comes towards light. Larvae feed on willows and apples. Overwinters as a pupa in the soil.

3 Poplar hawkmoth
Laothoe populi
SIZE AND DESCRIPTION Forewing is about 40 mm. Variable colouring, from grey to pinkish-brown. Orange patches on hindwings. At rest, the wavy-edged hindwings protrude beyond the leading edge of the forewings, giving the impression of a bunch of dead leaves. The green caterpillar, up to 60 mm long, has a yellow horn and seven diagonal yellow stripes.
HABITAT Woodland margins, river valleys and parks throughout Europe, except the far north.
FOOD/HABITS A slow-flying moth that is on the wing between May and September. Its larvae feed on poplars and sallows.

4 Privet hawkmoth
Sphinx ligustri
SIZE AND DESCRIPTION Forewing is up to 55 mm. Brown wings have black markings. There is a tan trailing edge to the forewing. The body is striped with pink and black. Green caterpillar has seven purple and white stripes on each side of its body.
HABITAT Woodland edges, hedges, parks and gardens across Europe, except Ireland, Scotland and the far north of Scandinavia.
FOOD/HABITS Flies June to July, drinking nectar on the wing, especially from honeysuckle. Larvae feed on privet, ash and lilac. Overwinters as a pupa in the soil.

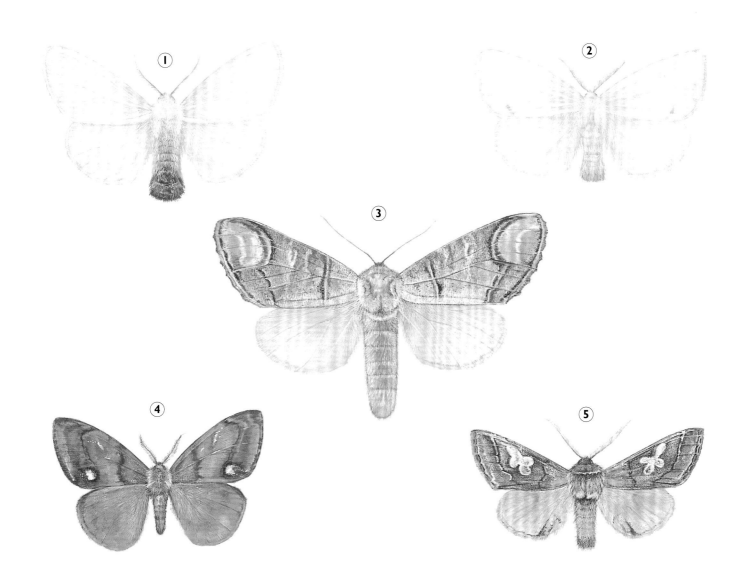

1 Brown-tail
Euproctis chrysorrhoea
SIZE AND DESCRIPTION
Forewing is up to 20 mm. Totally white and rather hairy. Males have brown abdomens. Females have white abdomens with brown tufts at the end. Caterpillar is black with a white stripe and yellow tufts of hair, which can cause rashes if touched. HABITAT Woods, hedges, parks and gardens from eastern and southern England across much of mainland Europe. FOOD/HABITS Flies July to August. Caterpillars feed in groups on many species of tree and bush.

2 Yellow-tail
Euproctis similis
SIZE AND DESCRIPTION
Forewing is up to 20 mm. Totally white and rather hairy. Males have thinner abdomens than females; both sexes have yellow tufts at the end. The black caterpillar is up to 40 mm long, with red stripes and white spots. It is very hairy, and contact with skin can cause a rash. HABITAT Woods, gardens, parks, orchards, tree-lined streets and hedges. Most of Europe, but rare in Scotland and Ireland. FOOD/HABITS Flies June to August. Larvae feed on hawthorn, blackthorn and fruit trees.

3 Buff-tip
Phalera bucephala
SIZE AND DESCRIPTION
Forewing is up to 30 mm. Silver-grey wings, with an orange tip to each forewing and an orange head, giving a broken-twig appearance when at rest. The caterpillar is about 45 mm long, with yellow stripes and sparse hairs. HABITAT Woods, parks, orchards and gardens throughout Europe. FOOD/HABITS Flies May to August. Caterpillars feed on leaves of oaks, limes, elms and other trees.

4 Vapourer
Orgyia antiqua
SIZE AND DESCRIPTION
Forewing is 15–18 mm. Chestnut, with a white spot on each forewing. Male is winged, but the female has only vestigial wings. Caterpillar is dark grey, with red spots and four cream tufts of hair on the back. Its body is covered with finer hairs. HABITAT Woods, parks, gardens, hedges and tree-lined streets across Europe. FOOD/HABITS Flies June to October, the males by day, the females at night. Caterpillars feed on a range of deciduous trees. Eggs overwinter.

5 Figure-of-eight
Diloba caeruleocephala
SIZE AND DESCRIPTION
Forewing is about 15 mm. Brown-and-grey forewing has a figure-of-eight marking. Hindwing is pale grey-brown. The grey-blue caterpillar has black spots and yellow lines. HABITAT Woodlands, scrub and gardens across Europe. FOOD/HABITS Flies September to October. Larvae feed on hawthorn, blackthorn and other rosaceous shrubs.

1 Common footman
Eilema lurideola
SIZE AND DESCRIPTION Forewing is about 15 mm. Pale grey forewing is fringed with yellow. Hindwings are pale yellow. Rests with nearly flat wings. The hairy grey larva has black lines on its back and red lines on its sides.
HABITAT Hedges, woods and orchards across Europe.
FOOD/HABITS Flies June to August. Larvae eat lichen.

2 Buff ermine
Spilosoma lutea
SIZE AND DESCRIPTION Forewing is 15–20 mm. Pale buff to creamy yellow, with a variable broken dark line on the forewing. The larva is up to 45 mm long and has tufts of long brown hairs.
HABITAT Most habitats, but especially common on waste ground and in gardens throughout Europe.
FOOD/HABITS Flies May to August. Larvae feed on wild and garden herbaceous plants.

3 White ermine
Spilosoma lubricipeda
SIZE AND DESCRIPTION Forewing is 15–20 mm. White, with more or less sparse white spots. Hairy thorax and yellow, black-spotted abdomen. The larva is up to 45 mm long, dark brown and very hairy, with a dark red line down its back.
HABITAT Hedgerows, gardens, waste ground and other habitats throughout Europe.
FOOD/HABITS Flies May to August. Adults do not feed, but larvae feed on herbaceous plants, including docks, dandelions and numerous garden plants.

4 Garden tiger
Arctia caja
SIZE AND DESCRIPTION Forewing is 25–35 mm. Chocolate-brown forewing has cream patterning. Hindwings are orange with black spots. The very hairy black-and-brown caterpillar is known as a "woolly bear".
HABITAT Open habitats, including gardens and scrub throughout Europe.
FOOD/HABITS Flies June-August. Caterpillar feeds on herbaceous plants. Winters as a small caterpillar.

Moths

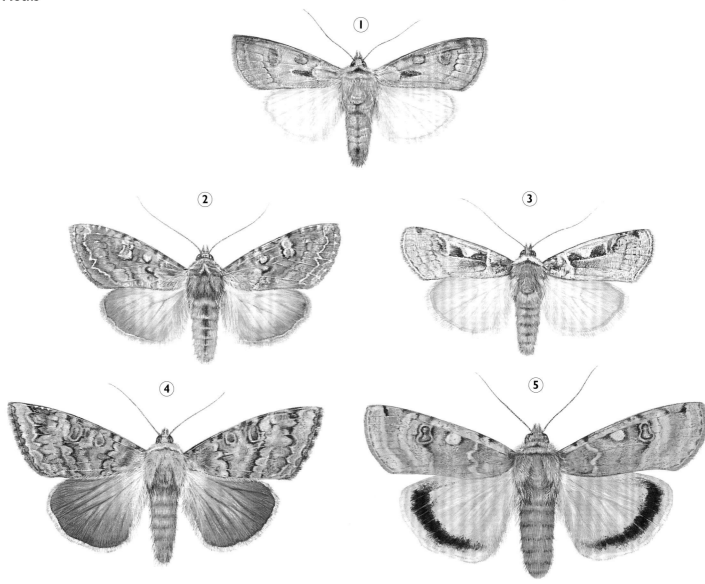

1 Heart and dart

Agrotis exclamationis
SIZE AND DESCRIPTION
Forewing is about 20 mm. Background colour varies from greyish-brown to deep brown. Wings have vaguely heart-shaped and dart-shaped markings. The larva, about 40 mm long, is a dull brown and grey "cutworm".
HABITAT Almost any habitat, especially cultivated. Found throughout Europe.
FOOD/HABITS Flies May to September at night. The larvae feed on the stems of herbaceous plants at night and hide in soil during the daytime.

2 Cabbage moth

Mamestra brassicae
SIZE AND DESCRIPTION
Forewing is 15–20 mm. Mottled greyish-brown, with rusty scales. The plump larva, up to 50 mm long, is brownish-green, with rather subtle dark and pale markings.
HABITAT Almost any habitat, but most common on cultivated land throughout Europe, except in the far north.
FOOD/HABITS Flies throughout the year, but mainly between May and September. The larvae feed on cabbages and other herbaceous plants. Winters as a pupa.

3 Setaceous hebrew character

Xestia c-nigrum
SIZE AND DESCRIPTION
Forewing is up to 20 mm. Greyish-brown to chestnut with a purplish tinge. There is a pale patch on the leading edge of the forewing. The larva is initially green, before becoming pale greenish-grey.
HABITAT Almost anywhere in Europe, except in the far north.
FOOD/HABITS Flies May to October, when it is at its commonest. The larvae have a diet of herbaceous plants. Passes the winter either in the larval stage or as a pupa.

4 Green arches

Anaplectoides prasina
SIZE AND DESCRIPTION
Forewing is about 20 mm. Greenish forewings have variable black markings, while the hindwings are dark grey or brown. The larva is brown with darker markings.
HABITAT Deciduous woodland over most of Europe.
FOOD/HABITS Flies mid-June to mid-July at night. The larvae feed on a range of plants, especially honeysuckle and bilberry.

5 Large yellow underwing

Noctua pronuba
SIZE AND DESCRIPTION
Forewing is 25 mm. Varies from pale to dark brown. The hindwings are deep yellow with a black border. The yellow flashes when the moth takes flight, which is thought to confuse predators. The green larva, up to 50 mm long, has two rows of dark markings on its back.
HABITAT Well-vegetated habitats across Europe, except in the far north.
FOOD/HABITS Flies June to October. Flight is fast and erratic. The yellow flashes shown in flight become invisible the moment it lands.

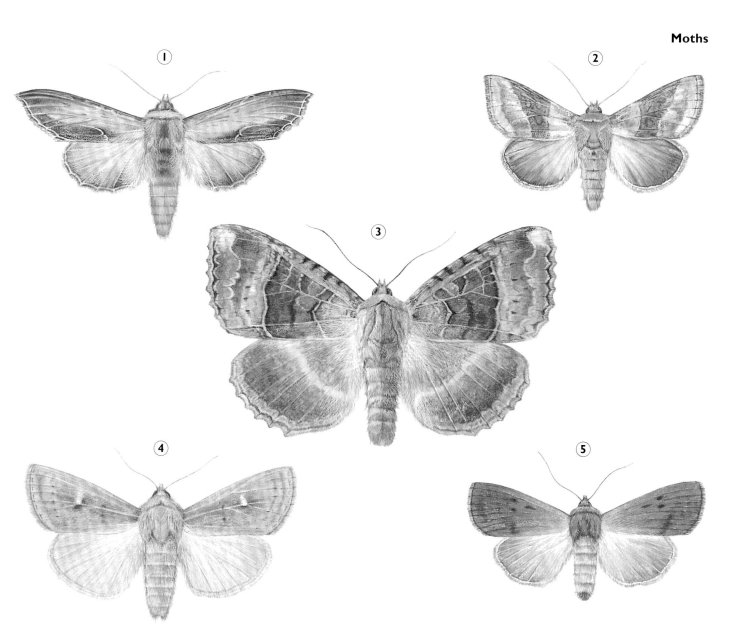

1 Mullein
Cucullia verbasci
SIZE AND DESCRIPTION
Forewing is 20–25 mm.
Colour varies from pale
straw to mid-brown and
is darker streaked. With
the wings held close to
its body, the resting moth
resembles a twig. The
larva, up to 60 mm long,
is creamy white and has
yellow and black spots.
HABITAT Woodland
edges, scrub, river banks,
gardens and parks over
most of Europe, but not
Scotland, Ireland and
northern Scandinavia.
FOOD/HABITS Flies April
to June. In June and
July, the larvae feed on
mulleins, figworts and
buddleia. The mullein
overwinters as a pupa.

2 Burnished brass
Diachrisia chrysitis
SIZE AND DESCRIPTION
Forewing is up to 20
mm. The two metallic
marks on the forewings
vary from emerald to
deep gold. Wings are
held above the abdomen
when at rest. There is
a prominent tuft on the
thorax. The larva, up to
35 mm long, is bluish-
green, with diagonal
white streaks across the
back and a white line
along the side.
HABITAT Gardens, parks,
hedges and waste ground
throughout Europe.
FOOD/HABITS Flies May
to October. Larvae feed
on nettles and mint.
Hibernates over winter in
its larval form.

3 Old lady
Mormo maura
SIZE AND DESCRIPTION
Forewing is 30–35 mm.
Patterned dark brown
and black, resembling
an old lady's shawl. The
greyish-brown larva, up
to 75 mm long, has dark
smudgy diamonds and a
broken white line running
down its back.
HABITAT Woods, hedges,
gardens, parks and damp
places in southern, central
and western Europe. Rare
in Ireland and Scotland.
FOOD/HABITS Flies July
and August, often coming
towards light. Larvae feed
on a variety of trees and
shrubs. Hibernates as a
small larva.

4 Clay
Mythimna ferrago
SIZE AND DESCRIPTION
Forewing is 15 mm.
Colour varies from straw
to reddish-brown, with
a white mark in the
middle of the forewing.
Wings lie flat when at
rest. The pale brown
larva is marked with
thin yellow lines.
HABITAT Common in
grassy places across
Europe.
FOOD/HABITS Flies May
to August. Larvae feed
on grasses and other low-
growing plants.

5 Mouse
*Amphipyra
tragopoginis*
SIZE AND DESCRIPTION
Forewing is 15 mm. Dark
brown, with three dark
spots. Underwings are
pale. Holds wings flat along
its abdomen when at rest.
The larva is green, with
narrow white lines.
HABITAT Widespread
in woods, hedgerows,
gardens and open
countryside with scrub
throughout Europe.
FOOD/HABITS Flies June to
September. Roosts by day
in outbuildings, under loose
bark and in hollow trees. If
disturbed, it scuttles off in
a mouse-like fashion. From
April to June, larvae feed on
plants such as salad burnet,
hawthorn and fennel.

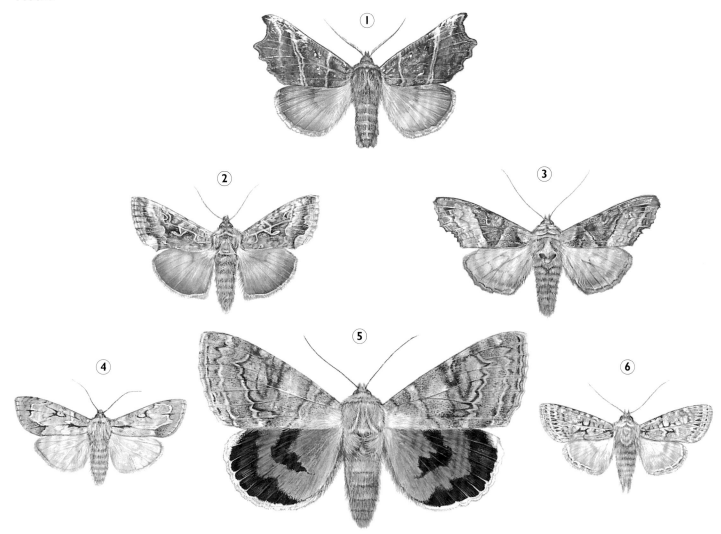

1 Herald

Scoliopteryx libatrix

SIZE AND DESCRIPTION Forewing is 20–25 mm. Purplish to orange brown, with bright orange scales near head. The trailing edge is ragged. The slender green larva is up to 55 mm long and has thin, pale yellow lines along its sides.

HABITAT Woodlands, gardens, parks and open countryside over Europe, except northern Scandinavia.

FOOD/HABITS Flies August to October, and in spring after migration. Larvae feed on willows and sallows. Overwinters as an adult.

2 Silver Y

Autographa gamma

SIZE AND DESCRIPTION Forewing is about 20 mm. Varies from purple-tinged grey to almost black, with a silver y-mark on the forewing. Green larva is up to 25 mm long.

HABITAT A migrant found all over Europe. Breeds all year in southern Europe. British and other northern breeders do not survive winter and are supplemented by migrants.

FOOD/HABITS Flies throughout the year. Attracted by nectar. May be seen in autumn alongside butterflies on buddleia. Larvae feed on low-growing wild and cultivated plants.

3 Angle shades

Phlogophora meticulosa

SIZE AND DESCRIPTION Forewing is about 25 mm. Varies from brown to green, with distinctive v-shaped markings. Forewing's trailing edge has a ragged look, exaggerated by its habit of resting with its wings curled over. The fat, green larva, up to 45 mm long, has a white line (often faint) along its back.

HABITAT A migrant found in almost any habitat in Europe.

FOOD/HABITS Flies most of year, but mainly May to October. Larvae feed on a variety of wild and cultivated plants. Overwinters as a larva.

4 Grey dagger

Acronicta psi

SIZE AND DESCRIPTION Forewing is 15–20 mm. Pale to dark grey, with dark, apparently dagger-shaped marks. The hairy, grey-black larva has a yellow line along its back, red spots along its sides and a black horn on its first abdominal segment.

HABITAT Woodlands, commons, parks and gardens across Europe, except the far north.

FOOD/HABITS Flies May to September, with larva feeding August to October on trees. Winters as a pupa.

5 Red underwing

Catocala nupta

SIZE AND DESCRIPTION Forewing is 30–35 mm. Grey mottled forewings make the moth well-camouflaged on tree-bark, but the bright red underwings are very conspicuous in flight. The pale brown caterpillar has warty, bud-like lumps on its back.

HABITAT Woodlands, hedges, gardens and parks across Europe, except northern Scandinavia.

FOOD/HABITS Flies August and September at night. Flies erratically, flashing the red underwings to confuse predators. Larvae feed from May to July on willow, poplar and aspen.

6 Grey chi

Antitype chi

SIZE AND DESCRIPTION Forewing is 15 mm. Mottled grey wings, with a small but distinct dark cross in the middle of the forewing. The bluish-green larva has green-edged white lines along its body.

HABITAT Gardens, grassy places and moorlands in much of Europe, but not the far north.

FOOD/HABITS Flies August and September, resting on walls and rocks during the day. Larvae feed from April to early June on low plants such as dock and sorrels. Overwinters as an egg.

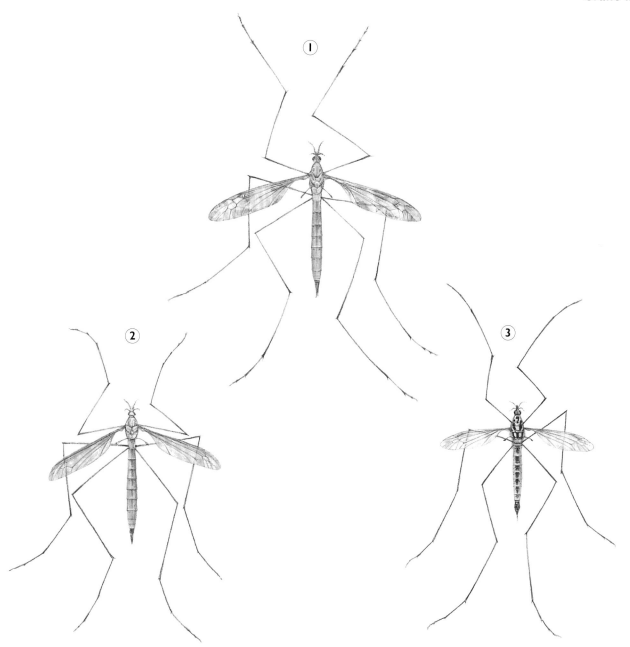

1 Large crane-fly
Tipula maxima
SIZE AND DESCRIPTION Almost 30 mm long, with mottled brown wings. Holds its wings at right angles when at rest. This is the largest of 300 British crane-fly species.
HABITAT Found in wooded areas, including gardens, across Europe.
FOOD/HABITS Adults fly April to August. Grubs live in the soil, feeding on roots.

2 Common crane-fly or daddy-long-legs
Tipula paludosa
SIZE AND DESCRIPTION About 25 mm long. Dark brown along the leading edges of the wings. The female's wings are shorter than its abdomen. The male has a square-ended abdomen, while the female's is pointed with an ovipositor. Dull brown grub is known as a "leather-jacket".
HABITAT Common in grasslands, parks and gardens throughout Europe.
FOOD/HABITS Flies throughout the year, but most numerous in autumn. Adults rarely feed. Grubs live in the soil and appear at night to gnaw the base of plant stems.

3 Spotted crane-fly
Nephrotoma appendiculata
SIZE AND DESCRIPTION 15–25 mm long, with a largely yellow abdomen, often with black spots. The female is larger than the male and has a pointed tip to the abdomen, while the male's abdomen is clubbed. Wings are clear and shiny. Fat, dark brown "leather-jacket" grub.
HABITAT Common in farmland, parks and gardens across Europe.
FOOD/HABITS Flies May to August. Adults rarely feed, but the grubs feed on roots and do considerable damage to garden plants.

Mosquitoes and midges

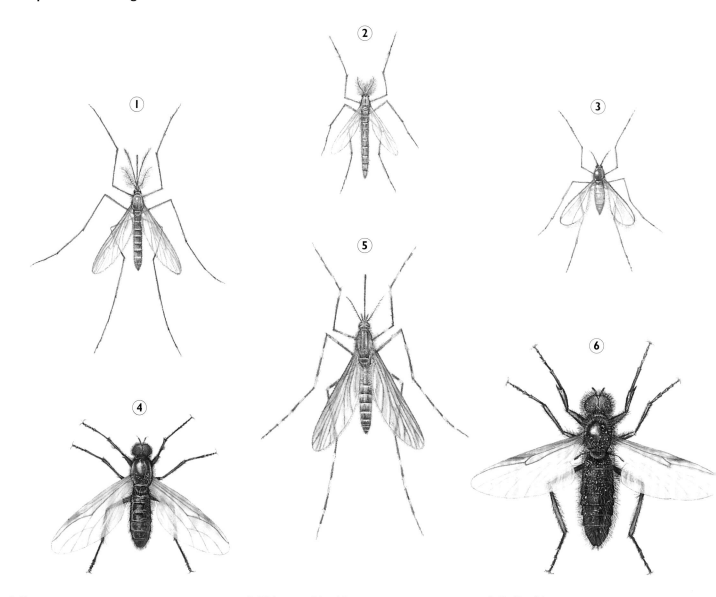

1 Common gnat
Culex pipiens
SIZE AND DESCRIPTION About 6 mm long.
Wings extend beyond abdomen's tip. Female
has a rounded tip to its abdomen. Male has
hairy antennae. Holds the abdomen parallel
to the surface on which it is perching. Aquatic
larvae live beneath the surface.
HABITAT Abundant throughout Europe.
FOOD/HABITS Flies at night with monotonous
hum. Rarely bites humans, preferring birds.
Eggs are laid in rafts on surface. Larvae dangle
beneath surface. Adults hibernate in sheds.

4 Fever-fly
Dilophus febrilis
SIZE AND DESCRIPTION About 4 mm long. Black,
but not hairy (compare St. Mark's fly). Female
has smoky, almost opaque wings, while male's
are almost clear with a black mark.
HABITAT Abundant in most open habitats.
FOOD/HABITS Flies from March to October,
but is commonest in spring. Large flocks may
cluster on grass stems. Males often hover
sluggishly. Larvae live beneath the soil in
decaying matter, but may damage roots.

2 Chironomid midge
Chironomus plumosus
SIZE AND DESCRIPTION About 8 mm long.
Wings are shorter than the abdomen and held
over the body at rest. Male's antennae are
very bushy. Reddish aquatic larva is known as a
"bloodworm".
HABITAT Must have a body of water in which
to lay eggs. This may be relatively small, such
as a water butt.
FOOD/HABITS Adults rest on walls as they dry
out after emerging from pupae. Non-biting.

5 *Theobaldia annulata*
SIZE AND DESCRIPTION About 6 mm long, this
is the largest mosquito in Europe. Has white
rings around the legs. Dark spots on the wings
are formed by convergences of veins.
HABITAT Widespread where there is stagnant
water for breeding.
FOOD/HABITS Adult females are blood-sucking
and require blood before they can lay fertile
eggs. Males feed on nectar and other plant
juices. Females hibernate in sheds.

3 Gall midge
Jaapiella veronicae
SIZE AND DESCRIPTION 2 mm long. Pointed tip
to abdomen, which is pale yellow. Wings are
hairy. Fine, bead-like antennae. Larvae are tiny
and orange.
HABITAT Open areas with small plants.
FOOD/HABITS Flies in swarms on summer
evenings, often entering lighted windows.
Grubs live inside germander speedwell plants
and create hairy galls on the tips of shoots.

6 St. Mark's fly
Bibio marci
SIZE AND DESCRIPTION 10–12 mm long. Black,
heavy-looking fly whose dangling, hairy legs add
to the impression of a lumbering flight. Larva is
primitive, with a large head.
HABITAT Gardens, woodland edges and well-
vegetated open country across Europe.
FOOD/HABITS Flies in late April and May (St.
Mark's day is 25 April). Suns itself on walls and
flowers. Larvae live beneath the soil, eating
rotting material.

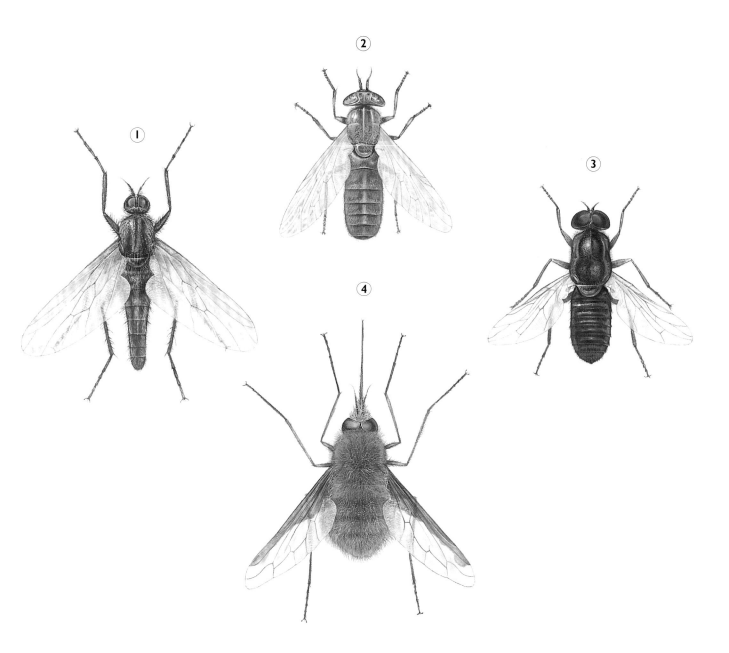

1 Dance fly
Empis tessellata
SIZE AND DESCRIPTION 10–12
mm long. Dark grey, with a small
head, pointed proboscis and
sturdy thorax. Long hairy legs.
HABITAT Woodland edges,
hedges, gardens and open
habitats with shrubs.
FOOD/HABITS Flies April to
August. Hops among hawthorn
or umbellifer flowers. Probes
blossom and hunts for other
insects, which it pierces with its
proboscis. May be seen on the
wing carrying flies it has caught.

2 Cleg-fly
Haematopota pluvialis
SIZE AND DESCRIPTION About 10
mm long. Dull grey, with a rather
cylindrical abdomen. The wings
are mottled. Holds its wings
above the abdomen when at rest.
Flies silently.
HABITAT Common from May
to September, especially in damp
woods. Replaced in north and
upland areas by another
similar species.
FOOD/HABITS Flies May to
October. Often seen in
thundery weather. Females are
bloodsuckers, biting human
beings and livestock. Males drink
nectar and plant juices. The larvae
live in the soil, where they prey
on other invertebrates.

3 Window fly
Scenopinus fenestratus
SIZE AND DESCRIPTION About
7 mm long. Small, black and
without bristles. Reddish-brown
legs, sometimes with black
markings. Wings are tightly folded
when at rest.
HABITAT Often seen at windows,
particularly in old buildings.
FOOD/HABITS Seems reluctant to
fly, apparently preferring to walk
away when disturbed. Larvae
live in birds' nests and buildings,
preying on other insects and their
larvae.

4 Large bee-fly
Bombylius major
SIZE AND DESCRIPTION
10–12 mm. Brown, furry, bee-
like coat and a long proboscis.
Dark leading edge to wings. Legs
are long and slender.
HABITAT Wooded places across
Europe, but rare in far north.
FOOD/HABITS Hovers, using its
long front legs to steady itself
as it reaches for nectar with
its long proboscis. The larvae
are parasitic on solitary bees and
wasps.

Hover-flies

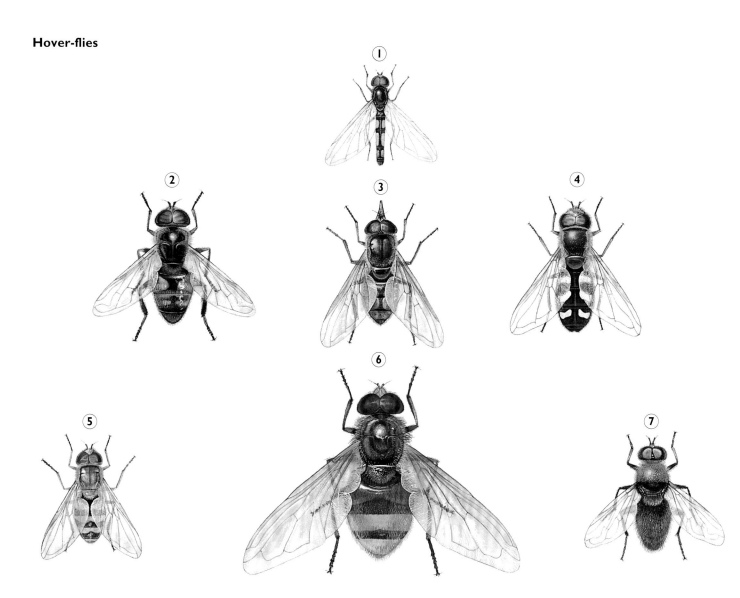

1 Hover-fly
Melanostoma scalare
SIZE AND DESCRIPTION 6–9 mm long. Narrow abdomen with yellow marks. Males abdomen narrower than females. Female's abdomen narrows towards thorax.
HABITAT Common in herb-rich areas and gardens.
FOOD/HABITS Flies between April and November. Often seen hawthorn blossom in May. Larvae feed on aphids.

3 Hover-fly
Rhingia campestris
SIZE AND DESCRIPTION About 10 mm long. Tan coloured with black marking down the center of abdomen with hairy fringe giving the impression of a rather bald bee. Long proboscis.
HABITAT Flower-rich habitats over most of Europe.
FOOD/HABITS Adults seen from April to November. Feeds on deep, tubular blue and purple flowers. Lays eggs in cow-dung.

5 Hover-fly
Syrphus ribesii
SIZE AND DESCRIPTION About 10 mm long. Yellow and black striped with rounded abdomen. Larva is green and slug-like. There are several similar species.
HABITAT Flower-rich habitats across Europe.
FOOD/HABITS Flying adults seen from March to November. Males perch on leaves or twigs up to 2.5 metres from the ground and make high-pitched whining noise. Feeds mainly on nectar, but will crush and swallow pollen. Larva feeds on aphids and is itself a victim of parasitic wasps.

2 Drone-fly
Eristalis tenax
SIZE AND DESCRIPTION 10–15 mm long. Looks like a honey bee drone. Dark anvil marks on abdomen. Larva is called a "rat-tailed maggot".
HABITAT Very common in parks, gardens and other flower-rich places across Europe.
FOOD/HABITS Nectar and pollen eater. Can be seen throughout the year. Larva lives in stagnant water, foul water, sewage, and dung-hills.

4 Hover-fly
Scaeva pyrastri
SIZE AND DESCRIPTION 12–15 mm long. Black abdomen with six bold cream crescents. Rounded abdomen. Slug-like larva.
HABITAT Flower-rich habitats across much of Europe.
FOOD/HABITS Adults seen flying from May to November, but most commonly seen in late summer. Feeds on nectar and honeydew. Larvae feed on aphids.

6 Hover-fly
Volucella zonaria
SIZE AND DESCRIPTION 15–25 mm long. Has a hornet-like appearance.
HABITAT Woodlands, gardens from Mediterranean to southern England.
FOOD/HABITS Flies May to November. Feeds on nectar and pollen.

7 Narcissus-fly
Merodon equestris
SIZE AND DESCRIPTION 10–15 mm long. Bumblebee mimic.
HABITAT Gardens, parks, woods and hedges across Europe.
FOOD/HABITS Flies March to August. Larva burrow down into bulbs.

1 Stilt-legged fly
Sepsis fulgens
SIZE AND DESCRIPTION
About 3 mm long. Black
with long legs. Brown
spots on wings.
HABITAT Common in
open country.
FOOD/HABITS Particularly
numerous in autumn,
when dense swarms of
hundreds of thousands
scurry over umbellifers.
Hibernate as adults.
Larvae breed in dung.

2 Fruit-fly
Drosophila funebris
SIZE AND DESCRIPTION
About 3 mm long. Small
dark brown fly. One of
many similar species.
HABITAT Widespread in
gardens, farms, orchards
and food factories.
FOOD/HABITS Attracted
by rotting fruit, vinegar,
wine and other fermenting
material. Commonest in
summer and autumn, but
present all year in food
and drink factories. Larvae
feed on decaying vegetable
matter.

3 Common carrot fly
Psila rosae
SIZE AND DESCRIPTION
About 4 mm long. Black
thorax and abdomen
with brown legs.
Creamy white grub.
HABITAT Gardens and
farmland.
FOOD/HABITS Lays eggs
in late spring near young
carrots. Larvae infest the
roots, often turning them
into empty shells.

4 Celery fly
Euleia heraclei
SIZE AND DESCRIPTION 6
mm long. Wings mottled
dark or reddish brown.
Body bulbous.
HABITAT Gardens and
open countryside where
umbellifers grow.
FOOD/HABITS Flies from
April to November.
Larvae eat leaves of
umbellifers from the
inside, causing brownish
mines. A serious pest of
celery and parsnip.

5 Picture-winged fly
Urophora cardui
SIZE AND DESCRIPTION
7 mm long. Transparent
wings are heavily mottled
brown, darker in males.
HABITAT Open country
where creeping thistles
are found.
FOOD/HABITS Flies May
to August. Eggs laid in
creeping thistles and
larvae cause hard, egg-
shaped galls. The galls
become buried in litter
over winter and become
softened by melting snow
or spring rain, which
stimulates pupation.

1 Louse-fly
Ornithomyia avicularia
SIZE AND DESCRIPTION About
5 mm long. Green legs. Winged.
HABITAT Found on woodland
birds, such as owls, pigeons and
thrushes. More frequently found
on younger birds, because adults
preen more effectively.
FOOD/HABITS Found between
June and October. Feeds on the
blood of its hosts.

2 Green parasitic fly
Gymnochaeta viridis
SIZE AND DESCRIPTION
10–12 mm long. Metallic green
with hairy brown eyes, abdomen
fringed with hairs and bristly legs.
HABITAT Woods, parks and
gardens.
FOOD/HABITS Found on plants
from March to July. Eggs are laid
on plants and the larvae, when
small, bore into the caterpillars
of moths.

3 Flesh-fly
Sarcophaga carnaria
SIZE AND DESCRIPTION 12–20 mm
long. Red eyes. Grey and black
chequered abdomen. Large feet.
HABITAT Often seen around
houses, but rarely indoors. Wide
range of habitats throughout
Europe.
FOOD/HABITS Active throughout
the year. Adult feed on nectar,
rotting carrion and dung. Females
are live-bearers. Maggots feed in
dung and carrion.

4 Swallow louse-fly
Crataerina hirundinis
SIZE AND DESCRIPTION About
5 mm long. Squat, brown.
Flightless with much reduced
wings.
HABITAT Nests of swallows and
martins.
FOOD/HABITS Found on
swallows and martins between
May and October. Pupae
overwinter in the nest and
emerge when the birds return
in spring.

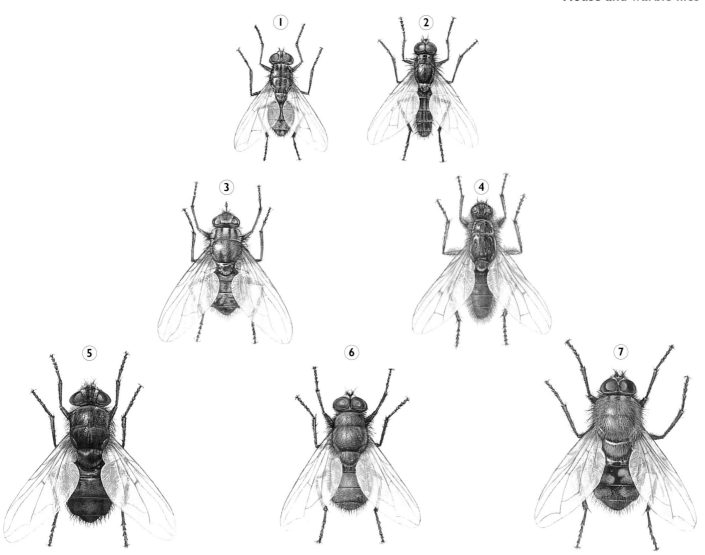

1 Common house-fly
Musca domestica
SIZE AND DESCRIPTION 8 mm long. Black and tan abdomen.
HABITAT In and around houses throughout Europe. Especially numerous in places where there is plenty of decaying matter.
FOOD/HABITS Found most of the year, but most common from June to September.

3 Stable fly
Stomoxys calcitrans
SIZE AND DESCRIPTION 8 mm long. Blackish abdomen rather short and round. Noticeable proboscis.
HABITAT Mainly around stables and farms throughout Europe.
FOOD/HABITS Flies from April to October. Adult sucks blood, using piercing proboscis. Bites horses, cattle and sometimes man. Larvae feeds in dung and stable litter.

5 Bluebottle
Calliphora vomitoria
SIZE AND DESCRIPTION 12–15 mm long. Rounded metallic blue body. Creamy white, carrot-shaped larva.
HABITAT Widespread through Europe. Often seen in and around houses.
FOOD/HABITS Seen all year round, often sunning on walls. Eggs are laid on meat and carrion, on which larvae feed.

2 Cabbage root fly
Delia radicum
SIZE AND DESCRIPTION 5–7 mm long. Bristly black or dark grey abdomen.
HABITAT All over Europe wherever there are crucifers growing.
FOOD/HABITS Flies March to November and feeds on nectar. Larvae feed on roots of brassicae and oilseed rape, causing leaves to become limp and yellow.

4 Yellow dung fly
Scathophaga stercoraria
SIZE AND DESCRIPTION 8 mm long. Males are covered with golden yellow. Females are grayish and less furry.
HABITAT Places where there is dung from horses and cows across Europe.
FOOD/HABITS Adults fly for much of the year, preying on other flies on cow-dung. Adults develop in cow-pats.

6 Greenbottle
Lucilia caesar
SIZE AND DESCRIPTION 8–15 mm long. Abdomen varies from blue-green to emerald and becoming coppery with age. Silvery below the eyes.
HABITAT Most habitats across Europe. Common around houses
FOOD/HABITS All seasons. Feed on nectar, carrion and wounds.

7 Cluster fly
Pollenia rudis
SIZE AND DESCRIPTION 8–10 mm long. Stocky black and grey abdomen
HABITAT Very common on grassland throughout Europe.
FOOD/HABITS Flies throughout the year. Feeds on pollen and nectar.

Wasps, bees, sawflies, ichneumons and ants

The huge order Hymenoptera has over 100,000 species world-wide. Typically a member of the order has two pairs of membranous wings, which are linked by a series of tiny hooks. The forewing is usually larger than the hindwing, which may be difficult to see when the wings are at rest. The head is large and is very mobile, it is connected to the thorax with a narrow neck. The antennae are often longer in the males than in the females. Mouth-parts in many species have toothed jaws to cope with solid food, but in other species they are adapted to feed on nectar. Bees, which are nectar-feeders, have long tubular tongues, but they also have biting jaws for nest-building. The eyes are compound and there are also three ocelli.

There are two sub-orders of Hymenoptera. One contains the sawflies and the other bees, ants, wasps and ichneumons. Unlike other hymenopterans, the sawflies have no noticeable waists and the abdomens are rather cylindrical at the front end. The females give the group its name, because they have saw-like ovipositors with which they cut slits in plants in order to lay their eggs. Most species are vegetarian, taking nectar and pollen from flowerheads, often from umbellifers. Some are carnivores which prey on other insects. All the larvae are vegetarians and look similar to the caterpillars of butterflies and moths, but they have at least six, as opposed to five, pairs of fleshy legs. Some species pupate inside the food plants, while others pupate in cocoons in the soil or leaf-litter, or attached to the food plant.

The remaining hymenopterans contains a huge variety of solitary, parasitic and social species. They all have the typical "wasp-waist", which is anatomically actually part of the abdomen but fused to the thorax.

The ichneumons tend to have narrower abdomens than the bees, wasps and ants, but they still have the narrow waist. Their forewings usually have a thickened vein along the leading edge. Almost all ichneumons are parasitic. The females have long ovipositors with which they pierce the host and lay their eggs within. Some have ovipositors long enough to pierce through the walls of plants to reach the grubs inside. Insects are the usual hosts, but some species attack spiders. The host is usually attacked at the early stages of the life cycle, when it is an egg, larva or pupa.

The ichneumon's larva grows inside its host and consumes it, eventually killing it and then pupating. It would, of course, be detrimental to the parasite if it killed its host too soon.

The ovipositors of bees, wasps and ants have become modified as a sting with which to stun prey or in defence. Bees are vegetarian, feeding on nectar and pollen, and their stings are purely defensive. Wasps are predators and the larvae almost always feed on animal material, but adults will take nectar and feed on the juices of rotting fruit. Ants feed in a variety of ways and species may be predators, vegetarians or omnivores.

All the ants are social insects, living in a colony centred upon a queen or queens. Males and females are winged and are seen in the dense, mating swarms in summer and early autumn. The females break off their wings after mating. The workers are asexual females and these are the individual ants that are mostly seen in gardens. Not all wasps or bees are social. There are many species of solitary bees and these often burrow into the ground to make their nests.

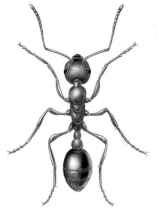

Digger wasps, which are social, have rather broad heads and spiny front legs with which they excavate nests in decaying wood. They feed their grubs with insects, which they have stunned and leave in the nest. Potter wasps are also social and build tiny nests of sand and mud. Solitary wasps include the spider-hunters. The tiny gall wasps lay eggs in plants. When the eggs hatch, the plant tissues around them swell to form galls. These come in several forms, such as spangle galls and robin's pincushions.

Key

The simple key below will help you to find out where to look when you are trying to identify wasps and their allies.

Ichneumons
A large family of Hymenoptera, whose larvae are parasites of other insects, particularly the larvae of butterflies and moths. Adults are slender with long antennae. Females of many species have long ovipositors.

Noticeable "waist"?
Narrow abdomen?
Long legs?
ICHNEUMONS page 63

Ruby-tailed wasps
Sometimes called cuckoo wasps, because they lay eggs in nests of solitary bees and wasps. Larvae feed on the bee or wasp larvae and pupae. Metallic coloured with hard exoskeleton which protects them from stings of host species.

Metallic colours?
RUBY-TAILED WASPS page 63

True wasps and Digger wasps
True wasps are social, eyes deeply notched or crescent-shaped. Digger wasps are solitary. Broad heads and an apparent collar on the front of the thorax. Females have spiny front legs for digging. Nest in rotten wood and hollow plant stems and mud.

Pointed yellow and black abdomen?
SOCIAL WASPS page 66
DIGGER WASPS page 66

Bees
Bees may be either social or solitary. Feed on pollen and nectar. Bodies are hairy. Pollen is often carried back to the nest on legs. Bees build their nests from wax from their own bodies. Honey bees sting in defence of their colonies, other bees will sting if handled.

Rounded abdomen?
BEES page 67

Bumble bees
Large, hairy and obvious. They are social, forming annual colonies, which break up in late summer. Only mated young queens survive the winter to form new colonies the next year. Nests are often established in disused rodent burrows and old birds' nests.

Rounded abdomen covered in hairs?
BUMBLE BEES page 69

Ants
Worldwide, there are about 15,000 ant species. Ants live in colonies that usually centre on a female, the queen, and are populated by males and workers. The workers are wingless females. Some ants use other species to do the work for them. Ants are "narrow-waisted", with a pedicel (the link between the thorax and abdomen) that has two segments. The antennae have a distinct "elbow" in them. Some species sting. Most species are omnivores that feed on nectar, seeds, the honeydew from aphids, and dead or living insects.

No wings?
ANTS page 65

Gall wasps
Gall wasps look like ants. Many species can fly. Eggs are laid in plants. When the eggs hatch the tissues around the grub swells to form a gall, which then provides the growing grub with food. The insects are less likely to be noticed than the galls that they produce.

Tiny with rather ant-like body?
GALL WASPS page 64

Sawflies
Although they belong to the same order as bees, wasps and ants, sawflies have no "waist". Many species have saw-like ovipositors with which females cut slits in plants before laying their eggs. Larvae look like moth caterpillars, but they have at least six pairs of legs (compared with the five pairs of moth caterpillars).

No waist?
Black and yellow with parallel-sided abdomen?
HORNTAIL page 62

Hairy thorax and abdomen?
HAWTHORN SAWFLY page 62

Sawflies

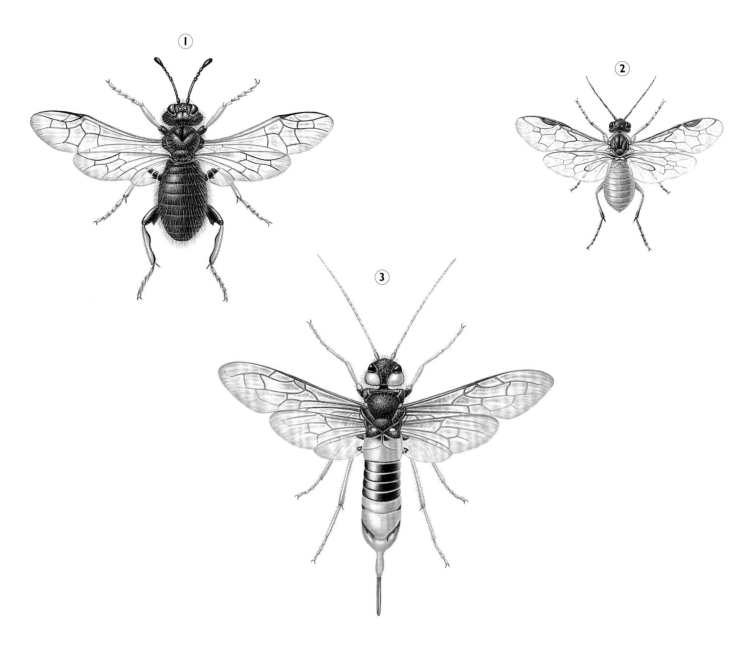

1 Hawthorn sawfly
Trichiosoma tibiale
SIZE AND DESCRIPTION 20 mm long. Leathery wings with hairy abdomen and thorax. Larva is pale green with brown head.
HABITAT Hedgerows, scrubby places and other habitats with hawthorn in northern and central Europe.
FOOD/HABITS Flies in May and June. Larva feeds on hawthorn and spins tough cocoon in which it pupates and from which the adult bites its way out.

2 Gooseberry sawfly
Nematus ribesii
SIZE AND DESCRIPTION Up to 10 mm. Female has yellow abdomen. The male's abdomen is black and is thinner. Green larva with black head.
HABITAT Common in gardens over most of Europe except the far north.
FOOD/HABITS Adults fly from April to September. Larvae feed gregariously on leaves of gooseberry and currants. Pupates in the soil.

3 Horntail
Urceros gigas
SIZE AND DESCRIPTION Up to 40 mm long, including the ovipositor. Female is black and yellow. The male is smaller with an orange abdomen with a black tip and orange legs.
HABITAT Coniferous woodland, but can survive in treated timber, from which they may appear in new houses.
FOOD/HABITS Despite the fearsome appearance of the females, horntails are harmless. Fly in sunshine from May to October. Males usually fly near tree tops. Females drill into bark and deposit eggs in the trunk. Larvae are almost legless and feed on the timber.

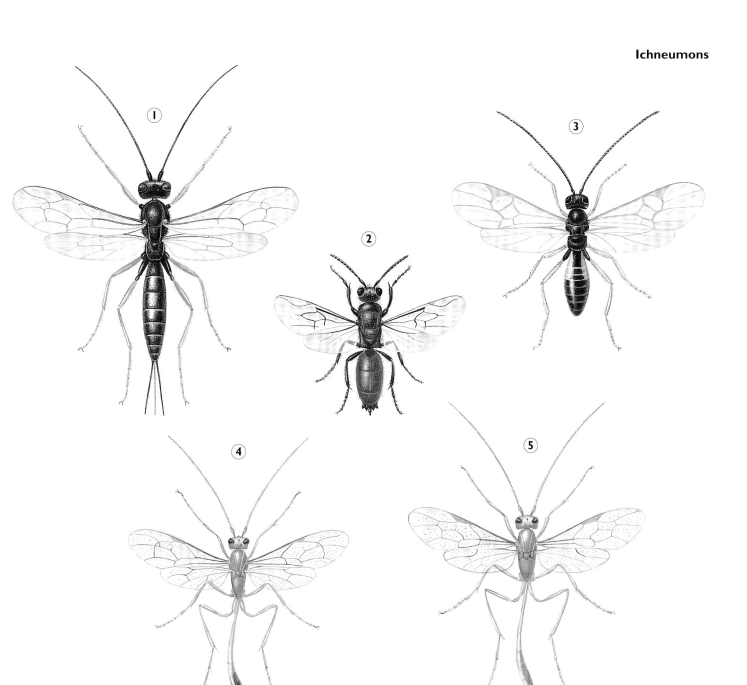

1 Ichnuemon fly
Pimpla instigator
SIZE AND DESCRIPTION
10–24 mm. Black body
with obvious orange legs.
The ovipostitor is roughly
half the length of the
abdomen.
HABITAT Most habitats
throughout Europe.
Present all summer.
FOOD/HABITS An
ectoparasite of moth
larvae, especially the
snout moth. The female
can inject as many as 150
eggs into a defenseless
caterpillar. The grubs
grow inside the caterpillar,
feeding on its body.

2 Ruby-tailed wasp
Chrysis ignita
SIZE AND DESCRIPTION
7–10 mm long. Abdomen
is brilliant red or purple.
Head is blue-green with a
golden sheen.
HABITAT A wide range of
open habitats throughout
Europe.
FOOD/HABITS Adults fly
from April to September,
feeding on nectar. May
be seen on walls and tree
trunks searching for nests
of mason wasps in which
to lay eggs. Larvae feed
on grubs of host and food
stored by host.

3 Ichneumon fly
Apanteles glomeratus
SIZE AND DESCRIPTION
3–4 mm long. Black with
smoky wings and brown
legs. Larvae are pale
brown, small and almost
translucent.
HABITAT Found
throughout Europe in
cultivated areas.
FOOD/HABITS Adults fly
in two broods in summer.
Eggs are laid in the
caterpillars of large white
and black-veined White.
Up to 150 grubs emerge
inside the caterpillar and
devour it to leave an
empty skin.

4 Ichneumon fly
Netelia testacea
SIZE AND DESCRIPTION
17 mm long. Yellowish
brown with strongly
arched abdomen with
dark tip.
HABITAT Found in well-
vegetated habitats through
much of Europe.
FOOD/HABITS Flies all
summer at night and
is attracted to lighted
windows. Feeds on the
host moth caterpillar from
the outside.

5 Yellow ophion
Ophion luteus
SIZE AND DESCRIPTION
15–20 mm. Yellowish
brown, strongly arched
abdomen and thorax.
Large black eyes.
HABITAT Well-vegetated
habitats throughout most
Europe except the far
north.
FOOD/HABITS Adults fly
from July to October.
Attracted by lighted
windows. Feeds on nectar
and pollen. Eggs are laid
in caterpillars or pupae of
several species. Usually one
grub per host. The adult
always emerges from the
host's pupa.

Gall wasps

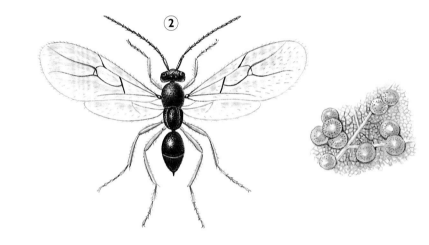

1 Oak apple gall
Biorhiza pallida
SIZE AND DESCRIPTION 1.7–2.8 mm. Small, brown wasp. Galls are spherical, reaching more than 30 mm in diameter.
HABITAT Oak trees.
FOOD/HABITS Oak apples appear on oaks in April or May. Males and females emerge from the galls in June or July. Males appear a day or two in advance of the females and the galls often contain only one sex. Females deposit eggs in the roots of the oak. Each gall is only about 10 mm in diameter and contains a single larva. The adults, which emerge in December or January, are all wingless females. These crawl up the tree to lay eggs in buds in the canopy. These result in oak apple galls and the sexual generation of gall wasps.

2 Spangle/Currant gall
Neuroterus quercusbaccarum
SIZE AND DESCRIPTION 75 mm long. Black with brown legs and antennae. Galls are brownish circles on underside of oak leaves, giving the impression of spangling or small and spherical like currants.
HABITAT Oak trees.
FOOD/HABITS Galls appear on leaves in late summer. In autumn the gall and its grub drop to the ground. The larvae pupate during the winter. The adults that emerge in February and March are all females which lay eggs on leaves and oak buds without fertilization. The larvae cause the formation of "currant" galls, which are small and spherical. Adult wasps, which emerge in May and June, are of both sexes. After mating the females lay eggs in leaves and the larvae cause "spangle" galls.

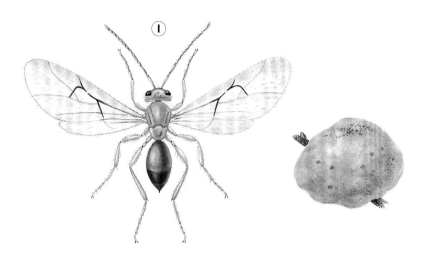

3 Robin's pincushion
Diplolepis rosae
SIZE AND DESCRIPTION About 4 mm long. Black head and thorax. Orange abdomen and legs. Reddish orange gall on wild roses. Whitish grub.
HABITAT Open countryside, woodland edges, gardens and parks with wild roses.
FOOD/HABITS Flies from April to June. Males are very rare. Females lay eggs without mating. Galls mature in autumn.

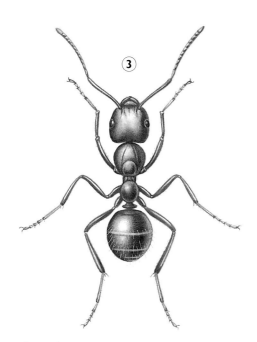

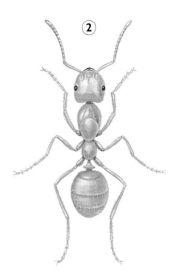

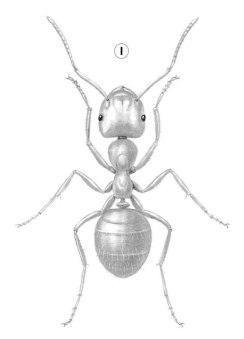

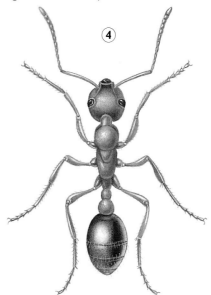

1 Yellow lawn ant
Lasius umbratus
SIZE AND DESCRIPTION Workers are 4–5 mm long and yellowish.
HABITAT Grassy places across Europe.
FOOD/HABITS Lives almost entirely below ground; usually only seen when the lawn is dug up.

2 Yellow meadow ant
Lasius flavus
SIZE AND DESCRIPTION Workers are up to 4 mm long. They are yellowish-brown, with a single-segmented pedicel. Males and females fly in July and August and are about twice the size of workers. The queen is darker. Yellow meadow ants do not sting.
HABITAT Orchards and rough grassland across Europe.
FOOD/HABITS Creates the anthills that are a characteristic feature of rough grassland. Mating flocks can be very numerous.

3 Black garden ant
Lasius niger
SIZE AND DESCRIPTION Workers are up to 5 mm long. They are black or dark brown, and the pedicel is a single segment. Flying ants, which emerge in July and August, are males and females. They are about twice the size of workers. Black garden ants do not sting.
HABITAT Open habitats throughout Europe, including gardens.
FOOD/HABITS Omnivorous, but is especially fond of sweet foods and will "milk" aphids for their honeydew. A colony consists of one queen and several thousand workers. Winged males and females emerge for mating flights in summer. Males die after mating; females break off their wings and seek a suitable nesting site. Birds take a heavy toll of the ants during mating flights. Very few females survive to create new colonies.

4 Red ant
Myrmica rubra
SIZE AND DESCRIPTION Workers are 4–5 mm long. They are chestnut brown, with a pedicel of two segments. Males and queens, which appear in late summer and early autumn, are about one-and-a-half times as long as workers. Males have longer, less bulbous abdomens than females. Red ants can sting.
HABITAT Open habitats throughout Europe.
FOOD/HABITS Omnivorous, and, although less inclined towards honeydew than the black ant, it tends towards animal food. A colony contains one or more queens and a few hundred workers.

Wasps

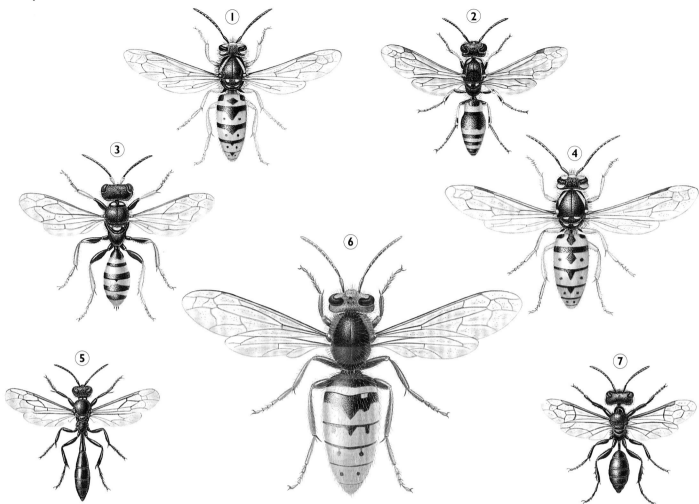

1 Common wasp
Vespula vulgaris
SIZE AND DECRIPTION Workers are 11–14 mm long. Black and yellow. Look for four yellow spots at the rear of the thorax. The yellow marks on either side of thorax usually have parallel sides.
HABITAT Common in most habitats across Europe.
FOOD/HABITS Usually nests in well-drained underground sites, such as hedgebanks, but will also use cavities in walls and lofts. Nests are built of yellowish paper.

3 Digger wasp
Ectemnius quadricinctus
SIZE AND DECRIPTION 13–14 mm long. Black-and-yellow abdomen, a large, squarish head and a narrow "waist".
HABITAT Woodland edges and gardens with wood-piles across central and southern Europe.
FOOD/HABITS Adults are visible from June to October. Feeds on pollen and catches flies on umbellifers. Nests in rotting wood, keeping a larder of paralysed insects, particularly hover-flies.

5 Black digger wasp
Trypoxylon figulus
SIZE AND DECRIPTION 6–12 mm long. Black, with a slender, tapering abdomen in which the segments are clearly visible.
HABITAT Woodland and gardens across Europe, but not northern Scotland and northern Scandinavia.
FOOD/HABITS Flies May to September. Nests in woodworm holes and hollow stems, stocking the brood cells with small spiders for the grubs to feed on.

2 Mason wasp
Ancistrocerus parietinus
SIZE AND DECRIPTION 10–14 mm long. Black and yellow, with a squarish black mark on the first yellow band of its pear-shaped abdomen.
HABITAT Common almost anywhere in Europe.
FOOD/HABITS Adults fly April to August. Feeds mainly on nectar and honeydew. The female makes a mud nest in a mortar cavity or natural crevice, and then stocks the nest with paralysed caterpillars.

4 German wasp
Vespula germanica
SIZE AND DECRIPTION Workers 12–16 mm. Looks very like the common wasp, but has marks on either side of the thorax bulge.
HABITAT Common in most habitats, except in northern Scandinavia.
FOOD/HABITS Nesting habits are similar to those of the common wasp, but the nest-paper is greyer and less brittle.

6 Hornet
Vespa crabro
SIZE AND DECRIPTION Workers 18–25 mm. Chestnut brown and yellow.
HABITAT Wooded areas, parks and gardens over most Europe, except Scotland, Ireland and northern Scandinavia.
FOOD/HABITS Nests in hollow trees, wall cavities and chimneys. Preys on insects as large as butterflies and dragonflies to feed young.

7 Digger wasp
Pemphredon lugubris
SIZE AND DECRIPTION 10–11 mm. Black, with a finely waisted abdomen.
HABITAT Woodland edges and gardens. Absent from northern Scotland and northern Scandinavia.
FOOD/HABITS Flies May to September. Nests in rotting wood, keeping a larder of aphids for the wasp larvae.

1 Flower bee
Anthophora plumipes
SIZE AND DESCRIPTION 14–16 mm long. Female is black, with hairy yellow-orange legs. Male has a rusty-brown thorax and a darkish tip to the abdomen. Looks like a bumblebee, but has a large eye that reaches the jaw.
HABITAT Many well-drained habitats. Common around human settlements. Much of Europe, but not Scotland.
FOOD/HABITS Flies from March to June. Feeds on nectar, using its long tongue to reach into tubular flowers. Nests in soil and soft mortar.

4 Tawny mining bee
Andrena fulva
SIZE AND DESCRIPTION 10–12 mm long. Female has a bright yellow abdomen, while the male, which is smaller, is dark.
HABITAT Open habits, including gardens, parks and woodland edges. Central and southern Europe, including southern England.
FOOD/HABITS Flies April to June. Nests in the ground, especially on lawns, throwing spoil from the nest hole into a small, volcano-like mound. Solitary species.

2 *Andrena hattorfiana*
SIZE AND DESCRIPTION 12–15 mm long. Dark brown with yellow towards tip of the abdomen. Largest British species.
HABITAT Anywhere in south and central Europe, including southern Britain, with flowers and soil in which to mine nesting tunnels.
FOOD/HABITS Flies from June to September. Female collects nectar and pollen in the summer to line the cells of her nest, she then lays an egg in each. Young hatch the following spring.

5 Mining bee
Andrena haemorrhoa
SIZE AND DESCRIPTION 10–12 mm long. Dark abdomen has a yellow tip, which is larger in the male. Female has a white face, while the male's face is pale brown.
HABITAT Woodland edges, scrub and gardens in northern and central Europe.
FOOD/HABITS An early spring species, which seeks nectar from blackthorn, sallow and dandelions. Solitary species.

3 Mason bee
Osmia rufa
SIZE AND DESCRIPTION 8–13 mm long. Black head and thorax, with a reddish-brown hair on the abdomen. The female is larger than the male, but the male has longer antennae. There are curved, bull-like horns between the female's antennae.
HABITAT Anywhere with flowers and suitable nest holes across central and southern Europe, including southern England.
FOOD/HABITS Flies April to July. Nests are in holes, and composed of several cells of mud.

6 Wool carder bee
Anthidium manicatum
SIZE AND DESCRIPTION 11 mm long. Thorax is black. Abdomen is black with yellow marks on either side. Legs are yellow. Not very hairy. Males are notably larger than females.
HABITAT Central and southern Europe.
FOOD/HABITS Flies June to August. Collects hairs from plants, which it carries in a ball beneath its body to take back to nest-holes in timber and masonry.

Bees

1 Cuckoo bee
Psithyrus campestris
SIZE AND DESCRIPTION 15–17 mm
long. The thorax has yellowish
hairs behind the head and grey
hairs in front of the abdomen,
which is largely shiny and
hairless. The male is smaller than
the female and varies in colour
from yellow to black. Does not
look like its host bumble bee, *B.
pascuorum*, which is unusual.
HABITAT Most habitats,
but avoids exposed places.
Throughout Europe, but
not Scotland and northern
Scandinavia.
FOOD/HABITS Parasitises
B. pascuorum.

2 Honey bee
Apis mellifera
SIZE AND DESCRIPTION 12–15 mm
long. Queens are about 20 mm
long, but are rarely seen outside
the nest. Colours vary. Can be
identified by the narrow cell near
the tip of the wing's leading edge.
Males have stouter bodies than
females.
HABITAT The honey bee is now
found almost everywhere.
FOOD/HABITS Flies spring to late
autumn. Lives in colonies with a
single queen. Males, or drones,
appear in spring and summer in
small numbers. Nests contain
combs of hexagonal cells, which
are used for rearing grubs and
storing pollen and honey.

3 Leaf-cutter bee
Megachile centuncularis
SIZE AND DESCRIPTION 10–12 mm
long. Dark coloured above, but
the female has an orange pollen
brush under the abdomen.
HABITAT Woods, gardens and
parks across Europe.
FOOD/HABITS Flies May to
August, visiting a range of
flowers. Female uses its jaws to
cut elliptical or round sections
from the leaves and petals of
roses and other plants. The leaf
pieces are then used to make
sausage-shaped cells for the
bee grubs.

4 Cuckoo bee
Psithyrus barbutellus
SIZE AND DESCRIPTION 20 mm
long. Resembles *Bombus hortorum*
(see page 69), but is less hairy.
The black abdomen has a white
tip.
HABITAT Gardens, parks and
open country across Europe.
FOOD/HABITS Parasitises *Bombus
hortorum* by laying eggs in the
nest, and often killing the queen.
The bumble bee workers then
rear the cuckoo bee's grubs as
if they had been laid by the
original queen.

1 Field bumble bee
Bombus pascuorum
SIZE AND DESCRIPTION
Up to 18 mm long. Can be identified by its reddish-brown thorax (darker in the northern part of its range). Thin covering of brownish hairs on abdomen.
HABITAT Well-vegetated habitats, but not in exposed places.
FOOD/HABITS Queens appear late March or April. Colonies live longer into autumn than other species. Nests in old birds' nests, nestboxes and long grasses at ground level.

2 Garden bumble bee
Bombus hortorum
SIZE AND DESCRIPTION
20–24 mm long. Collar, rear of thorax and first segment of abdomen are yellow. The tip of the abdomen is whitish. Scruffy appearance.
HABITAT Common in well-vegetated habitats, especially in gardens, throughout Europe.
FOOD/HABITS Queens often seen on white dead-nettle. Nests on or just beneath the ground.

3 Buff-tailed bumble bee
Bombus terrestris
SIZE AND DESCRIPTION
20–22 mm long. Orange collar and second abdominal segment. The tip of the abdomen is buffish-white; queen's abdominal tip is buffish in British Isles, but white elsewhere.
HABITAT Well-vegetated habitats across Europe.
FOOD/HABITS Queens visit sallow catkins in March and April; workers visit apple and cherry blossom. Nests well below ground level.

4 White-tailed bumble bee
Bombus lucorum
SIZE AND DESCRIPTION
20–22 mm long. Yellow collar and second abdominal segment, with white tip to abdomen.
HABITAT Well-vegetated places throughout Europe.
FOOD/HABITS A very early flier, with queens emerging in February and feeding on sallow catkins. Nests below ground.

5 Meadow bumble bee
Bombus pratorum
SIZE AND DESCRIPTION
16–18 mm long. Collar and second abdominal segment are yellow. Tip of the abdomen is orange-brown.
HABITAT Well-vegetated habitats across Europe, but not Scotland and northern Scandinavia.
FOOD/HABITS Appears in early spring, establishing colonies in April and May. Very agile, it visits both long, tubular flowers and open flowers. Nests on, below or above ground, including in nestboxes.

Beetles

The world's most numerous order of insects is the Coleoptera, which means "leather wings", and includes more than 300,000 species of beetles. Europe has over 20,000 species and there are 4000 species in Britain. The leathery casings, which usually cover the beetle's abdomen, are actually the insect's forewings, which are held vertical when it flies. The casings are known as elytra. Although most species of beetle can fly, they do not spend much time in the air. They live mainly on the ground, amongst leaf litter and vegetation, seeking refuge under stones and logs. Flight is used mainly for finding food and a mate.

Beetles have compound eyes and antennae which can be a useful aid to identification. Part of the head and the thorax are often protected by a horny shield known as a pronotum. The legs are variable and the variations can be a clue to the beetle's behaviour. Those that dig have toothed front legs and those that swim have flattened, paddle-like legs. The mouth-parts are adapted for biting, although some species are adapted to sucking plant juices or body fluids from other invertebrates.

The larvae of beetles are varied. Usually they have three pairs of legs, but in main the weevil larvae are legless. Water beetles have larvae that are aquatic, although the adults have the ability to fly and will fly from pond to pond. Water beetles carry the air they need underwater in the form of large bubbles between the elytra and the abdomen. Some smaller water beetles have adapted to avoid coming to the surface for air, by a gill-like mechanism for drawing oxygen from the water.

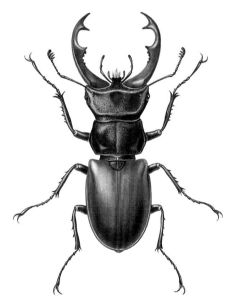

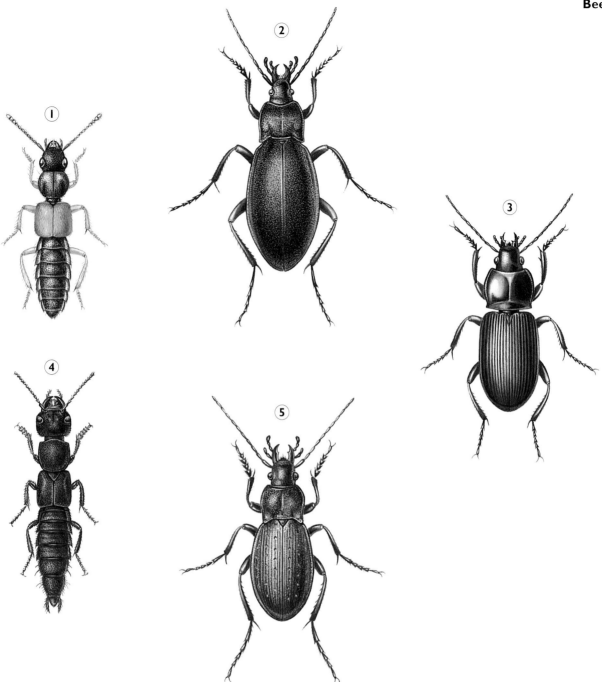

1 *Oxytelus laquaetus*
SIZE AND DESCRIPTION
6 mm long. One of the
many species of rove
beetle. There are over
a thousand in Europe.
Small brownish-orange
elytra and yellowish legs.
Black head, thorax and
abdomen.
HABITAT Compost heaps
in gardens.
FOOD/HABITS Feeds on
rotting vegetable material
and the grubs of other
insects in manure heaps.
Flies well.

2 Violet ground beetle
Carabus violaceus
SIZE AND DESCRIPTION
20–35 mm long. Black all
over, with violet tinges
to the thorax and elytra.
The thorax is flanged and
the elytra give a smooth
oval shape. Larva has
a shiny black head and
thorax, and a long, dusky,
segmented body.
HABITAT Woods, hedges,
gardens and scrub.
FOOD/HABITS Non-flying,
fast-running, nocturnal
predator of invertebrates.
Larva is also a predator,
but is less agile.

3 Black beetle
Feronia nigrita
SIZE AND DESCRIPTION
16 mm long. Jet black,
with ridges running down
the elytra.
HABITAT Woods,
gardens, and parks across
Europe.
FOOD/HABITS Nocturnal
predator of other
invertebrates.

4 Devil's coach horse
Staphylinus olens
SIZE AND DESCRIPTION
20–30 mm long. Black,
with small, almost square
elytra, which leave the
long abdomen exposed.
HABITAT Woods, hedges,
parks and gardens across
Europe. Often found in
damp outhouses.
FOOD/HABITS Nocturnal
predator with powerful
jaws. Feeds on slugs
and other invertebrates.
When under threat, the
beetle raises its tail and
opens its jaws.

5 Carabid beetle
Carabus nemoralis
SIZE AND DESCRIPTION
20–30 mm long. Black,
tinged with metallic
colours varying from
bronze to brassy green.
Elytra are pitted in lines
and finely ridged. Females
are less shiny than males.
HABITAT Most habitats
across Europe, except
northern Scandinavia.
FOOD/HABITS The carabid
is a fast-moving, flightless
beetle. It is a nocturnal
predator of ground-living
invertebrates.

Beetles

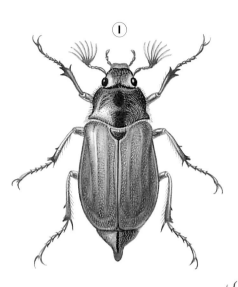

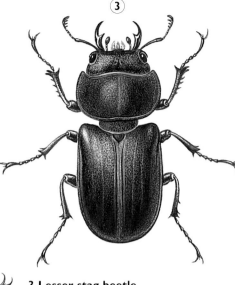

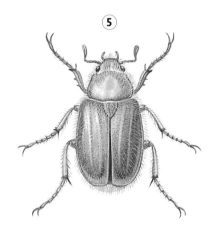

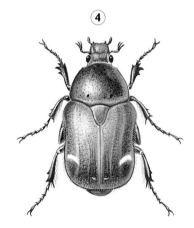

1 Cockchafer
Melolontha melolontha
SIZE AND DESCRIPTION 20–30 mm long. Black thorax. The rusty elytra do not quite cover the abdomen, exposing the pointed tip. Its legs are brown and the antennae fan out. Males have larger antennae than females. The whitish larva has a brown head, but is smaller and more wrinkly than the stag beetle larva.
HABITAT Woodland margins, parks and gardens. Common throughout Europe, but absent above 1,000 m and from northern Scandinavia.
FOOD/HABITS Also called the Maybug. Flies May to July at night. Adults chew leaves of trees and shrubs. Larvae, which take three years to develop, feed on roots.

4 Rose chafer
Cetonia aurata
SIZE AND DESCRIPTION 14–18 mm long. The flattened, squarish elytra are green, but may be bronze or bluish-black.
HABITAT Woodland margins, hedges, scrub and gardens in southern and central Europe, including southern England.
FOOD/HABITS Adults fly May to August by day and nibble the petals and stamens of flowers. Larvae feed in decaying wood.

2 Stag beetle
Lucanus cervus
SIZE AND DESCRIPTION 25–75 mm long. Smooth, dark-tan elytra, black head and thorax. Male's huge jaws look like antlers (hence the name). Whitish larva has a brown head.
HABITAT Oakwoods, parks and gardens. England, central and southern Europe. Becoming rare.
FOOD/HABITS Flies May to August, evenings and at night. Feeds on tree sap. Breeding males battle with "antlers". Larvae eat rotting wood.

3 Lesser stag beetle
Dorcas parallelipipedus
SIZE AND DESCRIPTION 19–32 mm long. Black and similar to the female stag beetle, but has only one spur on the tibia of its middle legs. Males have a particularly wide head. Larva is fat and whitish with a tan-brown head.
HABITAT Deciduous woods, parks and gardens in northern and central Europe.
FOOD/HABITS Flies April to October. Feeds on sap. Larvae live in rotting wood.

5 Summer chafer
Amphimallon solstitialis
SIZE AND DESCRIPTION 14–18 mm long. Entirely brown and hairy. Antennae have only three flaps to them.
HABITAT Parks, gardens, scrub and hedges across most of Europe.
FOOD/HABITS Adults swarm around deciduous trees and bushes at dusk and at night. Will fly to lighted windows. Larvae feed on grass roots and take up to two years to mature.

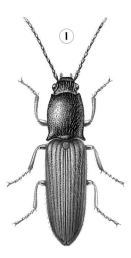

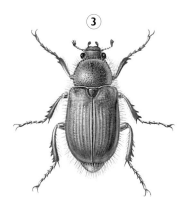

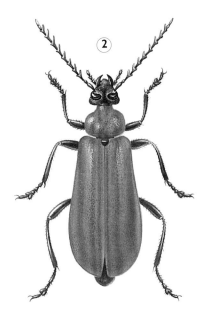

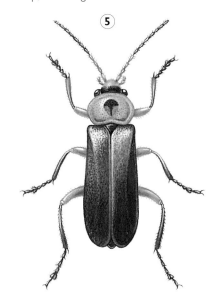

1 Click beetle
Athous haemorrhoidalis
SIZE AND DESCRIPTION 7–10 mm long. A long, black or dark brown thorax and a brown, ridged back. Larva is brown, with a thin, segmented body.
HABITAT Grassland, including parks and gardens, across Europe, except for northern Scandinavia.
FOOD/HABITS Flies May to July. Adults chew grasses and flowers, especially stamens with pollen. Larvae cause severe damage to roots. Click beetles, of which there are several species, are so-called because of their ability to flip themselves out of danger, making an audible "click" as it does so.

2 Cardinal beetle
Pyrochroa coccinea
SIZE AND DESCRIPTION 14–18 mm long. Bright reddish-orange elytra and thorax, with a black head and feathery antennae. Black legs. Larvae are yellowish brown with squarish rear ends.
HABITAT Woodland edges in central and northern Europe.
FOOD/HABITS Flies May to July. Found on flowers and old tree-trunks. Larvae live under bark and prey on other insects.

3 Garden chafer
Phyllopertha horticola
SIZE AND DESCRIPTION 8.5–11 mm long. Dark metallic green or black head and thorax, with brown elytra, sometimes with iridescence.
HABITAT Rough grassland, orchards and gardens with large lawns throughout Europe, except northern Scandinavia.
FOOD/HABITS Flies June and July, usually by day. Adults chew leaves of trees and shrubs. Larvae, which take two or three years to develop, feed on grass roots.

4 *Agriotes lineatus*
SIZE AND DESCRIPTION 7.5–10 mm long. Bullet-shaped, with lined elytra. Thorax is black or brown.
HABITAT Grassland and cultivated land in central and northern Europe.
FOOD/HABITS Seen throughout most of the year, but is commonest from May to August. The larvae, commonly known as "wireworms", can be a serious pest to cultivated plants.

5 Soldier beetle
Cantharis rustica
SIZE AND DESCRIPTION 11–14 mm long. Black elytra. The orange thorax bears a dark mark. Beaded antennae. Larva has flattened, segmented dark brown body with a pair of legs on each of the first three segments.
HABITAT Abundant throughout Europe in damp situations, including woodland edges and open country.
FOOD/HABITS Flies May to August. Preys on other insects, found on flower blooms.

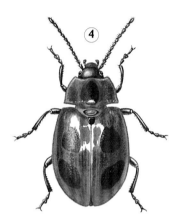
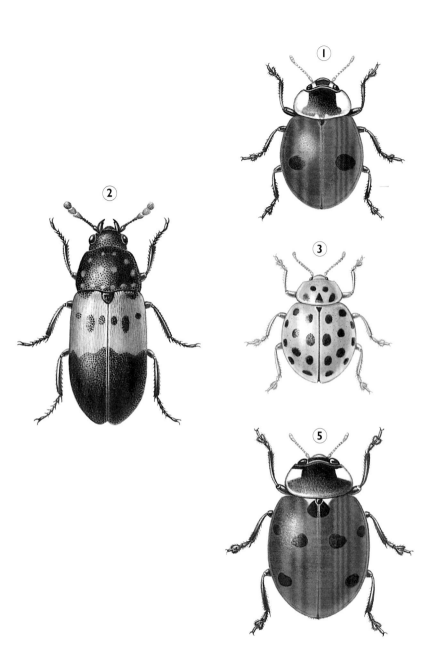

1 Two-spot ladybird
Adalia bipunctata
SIZE AND DESCRIPTION
3.5–5.5 mm long. Varies
greatly, with northern
populations often
being largely black. The
common form is red with
a bold black spot on each
elytron. Larva is similar to
the seven-spot.
HABITAT Well-vegetated
habitats across Europe.
FOOD/HABITS Flies spring
to autumn, eating aphids
on herbaceous and
woody plants. Winters in
groups (sometimes up
to a thousand) in sheds
and houses.

2 Larder beetle
Dermestes lardarius
SIZE AND DESCRIPTION
7–9.5 mm long. The larder
is an oval-shaped beetle.
The pale markings on the
elytra may be greenish,
greyish or brownish.
Larvae is short and
covered in hairs.
HABITAT Found in houses
and other buildings, and
also birds' nests. Central
and northern Europe.
FOOD/HABITS Adults are
found all year round, with
both larvae and adults
feeding on carrion and
dried meats in store. In
houses, the larvae feed
on animal products.

**3 Twenty-two-spot
ladybird**
Thea 22-punctata
SIZE AND DESCRIPTION
34.5 mm long. Lemon
yellow, with ten or eleven
black spots on each
elytron and five spots on
the pronotum.
HABITAT Well-vegetated
places across Europe.
FOOD/HABITS Flies April
to August. Eats mildew
on umbellifers and other
plants. Winters in leaf-
litter, but may appear in
mild weather.

4 *Endomychus coccineus*
SIZE AND DESCRIPTION
4–6 mm long. Red, with
four large black spots. Less
rotund than a ladybird.
Larva is brightly coloured.
HABITAT Woodland,
especially beechwoods.
FOOD/HABITS Flies April
to June. Eats fungus and
lives under the bark of
dying or dead trees.
The larvae crawl around
openly on wood fungi.

5 Seven-spot ladybird
Coccinella 7-punctata
SIZE AND DESCRIPTION
5.2–8 mm long. Bright-
red elytra, with seven
black spots. Larva is
steely blue, with yellow
or cream spots.
HABITAT Well-vegetated
habitats throughout
Europe. Abundant.
FOOD/HABITS Flies early
spring to autumn. Both
adults and larvae feed
on aphids. Winter is
passed in small groups or
individually in leaf-litter
and sheltered places near
to the ground.

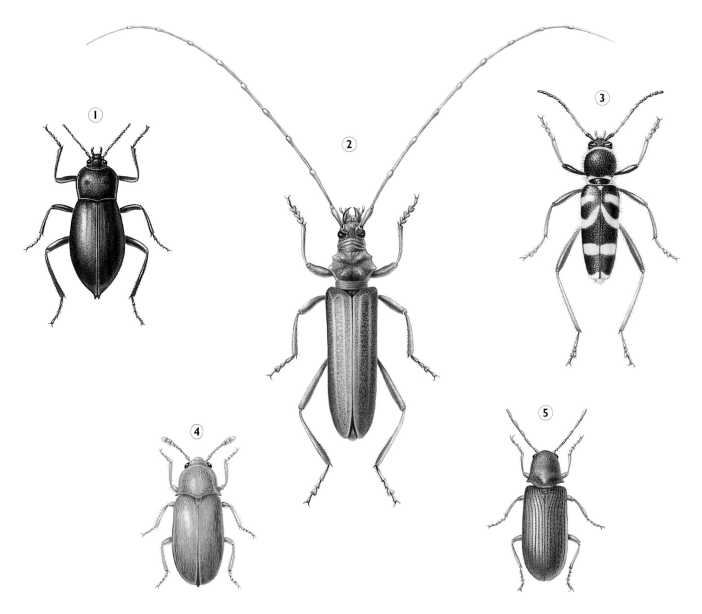

1 Churchyard beetle
Blaps mucronata
SIZE AND DESCRIPTION 18
mm long. Entirely black,
with pitted elytra that
taper at the ends to form
a pointed tip.
HABITAT Dark places
such as caves, cellars and
stables in northern and
central Europe.
FOOD/HABITS A
nocturnal, flightless
scavenger on vegetable
matter. Emits a foul smell
when threatened.

2 Musk beetle
Aromia moschata
SIZE AND DESCRIPTION
13–34 mm long. Striking
metallic green or blue.
Antennae as long as, or
longer than, the combined
head and body length.
HABITAT Deciduous
woodland especially
willow, across central
Europe. Local in southern
Britain.
FOOD/HABITS Flies June
to August. Emits a musky
secretion. The larvae
develop in willows,
particularly old pollards.

3 Wasp beetle
Clytus arietis
SIZE AND DESCRIPTION
7–14 mm long. Black, with
very variable yellow bands
on its elytra. Long-legged.
HABITAT Woods, gardens,
parks and hedges across
Europe, except northern
Scandinavia.
FOOD/HABITS Seen May
to July, often feeding on
flower nectar and pollen.
Female lays eggs in dead
wood. A harmless wasp-
mimic.

4 Raspberry beetle
Byturus tomentosus
SIZE AND DESCRIPTION
3.2–4 mm long. Yellowish-
brown or greyish and
hairy.
HABITAT Open areas
with scrub and bushes,
including gardens. Occurs
throughout central and
northern Europe.
FOOD/HABITS Adults,
which are present May
to July, gnaw flower
buds. The larvae develop
in blackberries and
raspberries, feeding on the
developing fruit.

5 Furniture beetle
Anobium punctatum
SIZE AND DESCRIPTION
2.5–5 mm long. Colour
of elytra varies from dark
brown to yellowish. Elytra
are ridged. Antennae are
clubbed. Covered with
fine down.
HABITAT Dry wood of
deciduous and coniferous
trees. Abundant in houses.
Central and northern
Europe.
FOOD/HABITS Seen
May to July. Larvae
are woodworm. Their
presence is shown only
by the escape holes of the
emerging adults, which are
1.5–2 mm in diameter.

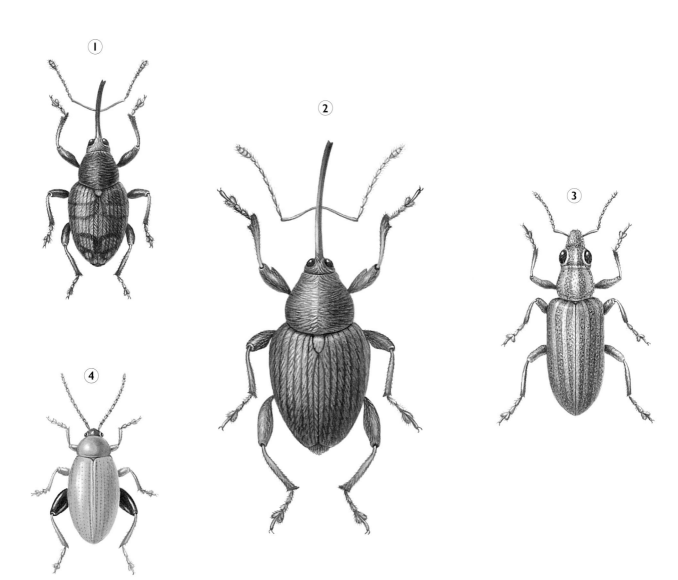

1 Oak weevil
Curculio villosus
SIZE AND DESCRIPTION 4–5 mm in length, including the "snout". Patchy grey and black.
HABITAT Woods, parks and gardens with oaks throughout Europe, except in the far north.
FOOD/HABITS It is found in oak trees in summer. The female uses her long snout to bore into acorns, where she lays her eggs. The larvae develop within the acorns.

2 Nut weevil
Curculio nucum
SIZE AND DESCRIPTION 6–9 mm in length, including the "snout". The snout, or rostrum, is longer in females than in males. Feathery antennae stem from the rostrum.
HABITAT Woods, parks and gardens with oak and hazel. Central and northern Europe.
FOOD/HABITS Adults seen April to July visiting hawthorn blossom for nectar. The female uses her long snout to drill into a young hazel nut, and then lays an egg in the hole. The emerging larva feeds on the kernel until autumn, when the nut falls to the ground, The larva then gnaws its way out of the nut and digs into the soil to pupate over winter.

3 Pea weevil
Sitona lineatus
SIZE AND DESCRIPTION 4–5 mm long. Pale and dark brown stripes run along the body. The eyes are very prominent.
HABITAT Found wherever wild and cultivated leguminous plants grow. Absent from northern Scandinavia.
FOOD/HABITS Adults, which are mainly active in spring and autumn, chew semi-circular pieces from the edges of leaves and may damage seedlings. The larvae live inside root nodules. There are several species of weevil that attack garden plants.

4 Potato flea beetle
Psylliodes affinis
SIZE AND DESCRIPTION 2.8 mm long. A reddish-brown beetle, with thick black thighs on its hind legs.
HABITAT Common on nightshades and potatoes in continental Europe.
FOOD/HABITS Adults nibble leaves, while larvae feed on roots of nightshades and potatoes. Can be a serious pest, but rarely causes problems in the British Isles.

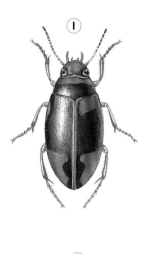

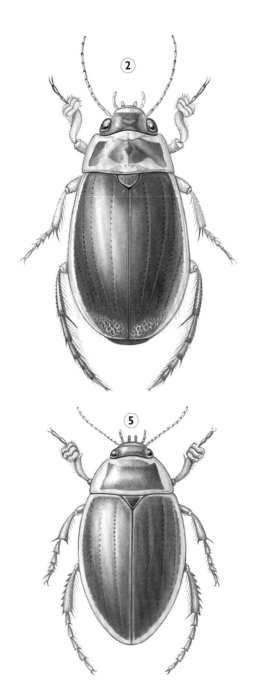

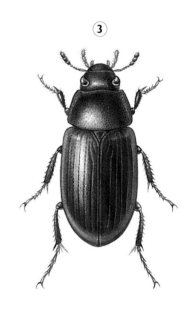

1 Hydroporus palustris
SIZE AND DESCRIPTION
3–3.3 mm long. Oval-shaped elytra. Black, with patches of orange or yellowish-brown. One of 34 species in a genus found in Europe.
HABITAT Common in all types of still water throughout Europe.
FOOD/HABITS Preys on a variety of small aquatic invertebrates. Flies at night. Comes to the surface for air, which it carries beneath its elytra in a bubble when it dives.

2 Great diving beetle
Dytiscus marginalis
SIZE AND DESCRIPTION
27–35 mm long. Very dark brown to black, fringed with yellowish-brown. Females have ridged elytra and males smooth. Larva has well-developed legs and a segmented body that bends more than a dragonfly nymph.
HABITAT The commonest European diving beetle. Prefers reedy ponds and other still waters.
FOOD/HABITS Flies at night. Preys on newts, tadpoles, fish and invertebrates. Larvae live underwater and are voracious predators.

3 Hydrobius fuscipes
SIZE AND DESCRIPTION 6 mm long. Black, with a metallic sheen. Can be distinguished by the pitted furrows along the elytra. The legs are rust coloured. The larva is maggot-like.
HABITAT Still waters across Europe.
FOOD/HABITS Omnivorous scavenger that does not swim well and crawls over underwater plants. Collects air from the surface by swimming to the surface head-first and storing air beneath its elytra. Larvae are carnivores.

4 Whirligig beetle
Gyrinus natator
SIZE AND DESCRIPTION
6–7 mm long. Tiny, shiny black beetle that gyrates on the water's surface. Middle and hind legs are short and oar-like. Two-part eyes enable it to look down into the water and across the surface simultaneously.
HABITAT Still and slow-moving water. There are several European species.
FOOD/HABITS Visible for much of the year, but hibernates. Preys on mosquito larvae and insects that fall into the water. Often seen in small groups. Dives if alarmed.

5 Colymbetes fuscus
SIZE AND DESCRIPTION
17–19 mm long. Rather like a small great diving beetle, but smoother and more uniformly oval in shape. Often has a green iridescence.
HABITAT Stagnant, weedy ponds and ditches throughout central and northern Europe.
FOOD/HABITS Can be found for most of the year, but may hibernate during the coldest months. It is a predator of other invertebrates.

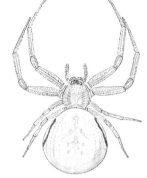

Spiders and harvestmen

These predatory invertebrates cause huge fear among some human beings, but in northern Europe they are harmless. They are certainly worth looking at instead of just throwing them out the window (or, worse still, stamping on them). Arachnida, the class that includes spiders, also includes harvestmen (page 83), scorpions, ticks and pseudoscorpions. Like insects, they have exoskeletons made of chitin, but they have four pairs of legs and their bodies are in two parts – the abdomen, which is soft and contains the digestive and sexual organs, and the cephalothorax, which is hard and contains the mouthparts and eyes and to which the legs are attached. The two parts of the body are connected by a narrow pedicel.

The legs of a spider are complex. They have seven segments and contain a fluid and muscle. They are moved by a combination of muscle contraction and hydraulic pressure. The eyes are simple compared with those of insects, and there are usually eight (but six in some species) Spiders are able to focus on objects and are sensitive to light and colour. In addition they have a sense of touch and can detect vibrations, air currents and tiny movements. They can also communicate chemically, detecting changes in pheromones and humidity.

The mouthparts are comprised of a labium, on either side of which are two maxillae (or jaws), and a pair of chelicerae, which consist of a basal portion, a fang and "teeth" which are used for biting, injecting venom and crunching prey. Behind the chelicerae are palps, which in males are modified as secondary sexual characteristics.
Abdomens vary in shape, size and patterning. The male's abdomen are smaller than female's. At the rear end of the abdomen are three pairs of spinners. Silk is produced from glands behind these and squeezed through the spinners. The silk is used for many functions other than making webs. Silk is used for the protection of egg sacs, for shelter and for travelling on air currents.

Eggs are laid in a silken sac or cocoon and may lay dormant over winter. The female of some species guard their eggs very carefully. The embryo develops into a prelarva, an incompletely developed spiderling which forces its way out of the egg membranes, but moults into the larval stage while still in the sac. The larva grows and moults again before emerging as a fully mobile spiderling, which is a perfect miniature of an adult except for its lack of sexual organs. Initially the young stay with the female. Females may overpower prey that is too large for the young to tackle or they may regurgitate food. Some species may disperse by drifting on air currents. They move to the end of a branch or leaves and create a strand of silk which catches the air current.

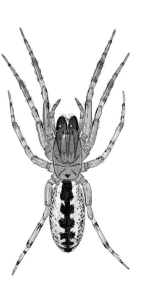

Spiderlings will moult five to ten times before adulthood. Adults males stop moulting, but some females continue. Moulting takes considerable effort, as the spider needs to force its way out of the exoskeleton. They are vulnerable at this stage as they must wait while the outer covering hardens. Some will die during the moult. If you find hollow remains of spiders they may be moulted exoskeletons rather than dead spiders.

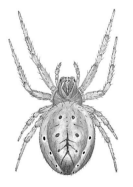

Males emerge from the final moult and seeks a female by detecting her pheromones. He may also attempt to attract a female by vibrating webs. The male transfers sperm from abdomen to his palps and may spin a tiny web on which the sperm is deposited. The male's approach to the female needs to follow the correct ritual if she is not to reject his advances.

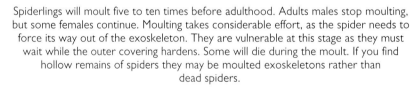

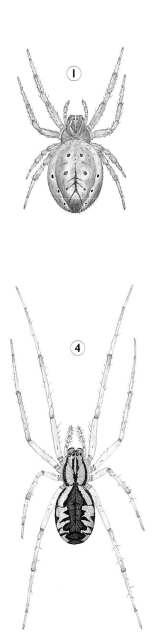

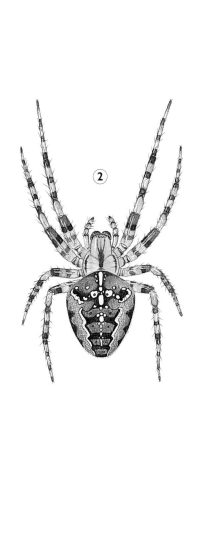

1 Green orb-weaver
Araniella cucurbitina
SIZE AND DESCRIPTION
3.5–6 mm long. Females
have a green abdomen
with dark brown spots
and a brown head and
legs. The slightly smaller
male has a smaller
abdomen and an orange-
brown head.
HABITAT Low bushes and
trees across Europe.
FOOD/HABITS Adults seen
summer to autumn. Small,
haphazard orb-web. Egg-
sacs are attached to the
underside of leaves and
covered by a mass of silk.
Females die in autumn.
Leaves bearing egg-sacs
fall to the ground; young
spiders emerge in spring.

2 Garden spider
Araneus diadematus
SIZE AND DESCRIPTION
Female is 10–13 mm long;
the male 4–8 mm. The
abdomen bears a white
cross. Male has a smaller
abdomen than the female.
Colours vary from pale
yellowish-brown to very
dark brown.
HABITAT Common in
woodland, heathland,
gardens and hedges
across northern Europe.
FOOD/HABITS A web-
spinner that preys on
flies and other insects. In
autumn, the female lays
up to 800 eggs in a single
mass, which are protected
by a layer of silk. She
stays with them until her
death a month later.

3 *Meta segmentata*
SIZE AND DESCRIPTION
4–8 mm long. Colours
are very variable, but the
pattern on abdomen is
more or less constant.
HABITAT Very common
throughout northern
Europe in gardens and
other well-vegetated
habits that will support its
orb-web.
FOOD/HABITS Adults are
mature in late summer
and autumn. When
disturbed, males especially
stretch their legs forward
along leaves or stems.
Webs are slung from
vegetation up to 2 m over
the ground. Spherical
egg-sacs are attached to
vegetation near the nest.

4 Hammock sheet-weaver
Linyphia triangularis
SIZE AND DESCRIPTION
5–6.6 mm long. Female
has an abdomen that is
roughly triangular in profile
and pale with brown
triangular marks down the
centre. Male's abdomen
is slimmer and lacks the
triangular marks.
HABITAT Widespread in
Europe wherever there
are trees or plants with
stiff foliage.
FOOD/HABITS Adults
seen mid-summer to
late autumn. Slings a
hammock-like web
in bushes, then hangs
beneath the web and
waits for insects to fall
into it.

5 Missing sector orb-weaver
Zygiella x-notata
SIZE AND DESCRIPTION
Female is 6–7 mm long;
the male 3.5–5 mm.
There is a dark, leaf-like
pattern fringed with pink
on the abdomen. Very
long front legs. The male
is similar to the female,
but smaller.
HABITAT Widespread
throughout Europe,
except Finland. Favours
human habitation.
FOOD/HABITS Slings a
vertical web around
window- and door-frames.
There are empty sectors
at top of the web. Spider
waits in a crevice for
insect prey to become
trapped in the web.

Spiders

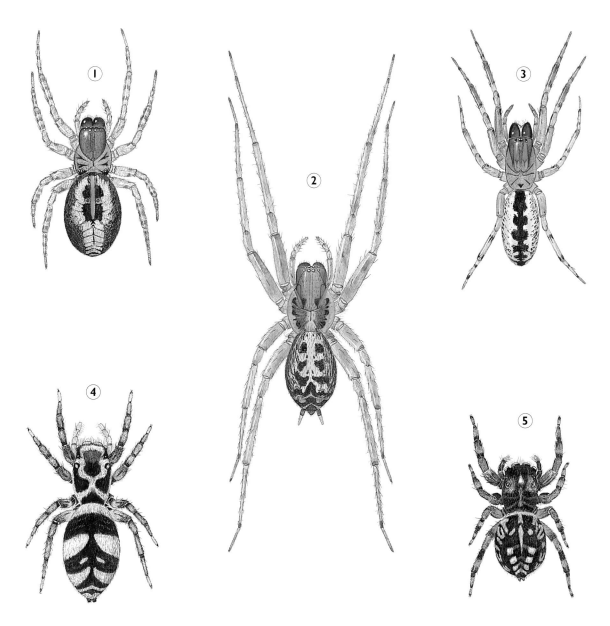

1 Mesh-web spider
Amaurobius similis
SIZE AND DESCRIPTION
Female is 9–12 mm
long; the male 6–8 mm.
Abdomen colours vary
from green to brown.
Dark marks on abdomen.
HABITAT Common and
widespread near human
habitation across Europe.
FOOD/HABITS Females are
found throughout most
of the year, but males are
seen only in late summer
and autumn. Spins a lace-
like web across a small
hole or crevice in which
the spider hides. Webs
tend to have a scruffy,
patchy appearance.

2 House spider
Tegenaria duellica
SIZE AND DESCRIPTION
11–16 mm long. Colour
is dark brown, with pale
markings. The male is
smaller than the female,
but has longer legs.
HABITAT Widespread in
northern Europe near
human habitation and in
rocky and wooded places.
FOOD/HABITS May be seen
running across floors at
night, especially in autumn,
when males are seeking
mates. Builds a triangular
web with a tubular retreat
in the corner, where the
spider waits for its prey
to become entrapped.
Females may live for
many years.

3 Leopard spider
Segestria senoculata
SIZE AND DESCRIPTION
7–10 mm long. Black
head but pale legs and
abdomen. The males
resemble females, but
have smaller abdomens.
HABITAT Holes in walls
and bark throughout
Europe.
FOOD/HABITS Adults
found spring to autumn.
Hides within holes, from
which about a dozen silky
trip-wires spread out.
When prey disturbs the
threads, the spider dashes
out of its hole to grab it.

4 Zebra spider
Salticus scenicus
SIZE AND DESCRIPTION
5–7 mm long. The male is
smaller than the female.
Black, with variable white
marks (hence the name)
and greyish legs. Hairy.
Short front legs and very
large eyes.
HABITAT Widespread
throughout northern
Europe. Often found on
walls and fences near
human habitation.
FOOD/HABITS Adults are
evident May to August.
Stalks prey, using its keen
eyesight, and then leaps
upon the victim. Active in
warm weather, especially
in sunshine.

5 Downy jumper
Sitticus pubescens
SIZE AND DESCRIPTION
4–5 mm long. Colour is
dull brown, but with light
patches and covered with
light hairs. Both sexes are
similar, but the male has a
smaller abdomen than the
female.
HABITAT Usually near
human habitation. The
spider is widespread,
but localized across
northern Europe.
FOOD/HABITS Hunts prey
by stalking and leaping
upon it. Jumps well.

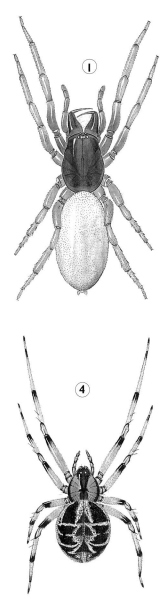

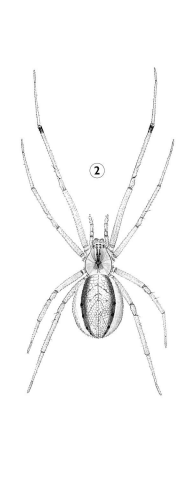

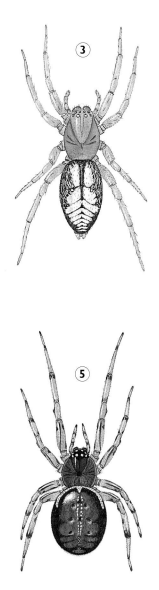

1 Woodlouse spider
Dysdera crocata
SIZE AND DESCRIPTION
9–15 mm long. A rather
fleshy-brown thorax and
legs. Abdomen is whitish.
Male are slightly smaller
than females, and have
narrower abdomens.
HABITAT Widespread
in Europe, except
Scandinavia. Found under
stones, logs and other
material in gardens and
slightly damp habitats.
FOOD/HABITS Feeds on
woodlice, which it catches
and crushes with its
fangs. Nocturnal. Spends
the day under cover in
a silken cell, in which its
eggs are laid. Found all
year round.

**2 Red-and-white
cobweb weaver**
Enoplognatha ovata
SIZE AND DESCRIPTION
3–6 mm long. Males are
smaller than females and
have smaller abdomens.
Very pale brown thorax.
Abdomen is creamy, with
two pink bands (as above),
a single broad pink band or
no band, but always with
pairs of black dots.
HABITAT Low vegetation
and bushes.
FOOD/HABITS Flimsy,
three- dimensional web
has sticky outer sections to
trap small insects. Female
guards her bluish egg-sac
beneath a leaf, which is
often rolled. Maturity is
reached in summer.

3 Fillet sac spider
Clubiona comta
SIZE AND DESCRIPTION
3–6 mm long. Males
are slightly smaller than
females. Pale brown
thorax and legs. Brown
abdomen with creamy
markings.
HABITAT Common
throughout northern
Europe in any habitat
with trees and bushes.
FOOD/HABITS A nocturnal
hunter that spends the
daytime hidden in silken
cells under stones,
amongst vegetation or
under bark. Found in
spring and summer.

4 Comb-footed spider
Theridion sisyphium
SIZE AND DESCRIPTION
2.5–3 mm long. Female
is larger than the male.
Brown thorax and a boldly
marked abdomen.
HABITAT Widespread
in northern Europe in
woodland margins, scrub,
hedges and gardens.
FOOD/HABITS Female spins
a three-dimensional web
of criss-cross strands on
bushes, particularly gorse.
She makes a retreat at
the top, where she rears
her brood. She guards
her greenish-blue egg-sac
and feeds her young by
regurgitation. Maturity is
reached in summer.

5 Cellar spider
Steatodea bipunctata
SIZE AND DESCRIPTION
4–7 mm long. Female
has a reddish-brown
abdomen with thin, pale
markings near the thorax.
The male is smaller, with
a narrow white marking
down the centre of the
abdomen and much larger
palps.
HABITAT Widespread and
common around houses
in northern Europe.
FOOD/HABITS Females are
found all year; males only
in summer and autumn.
Males have ridges and
teeth under the carapace
and abdomen, with which
they create sounds to
attract females.

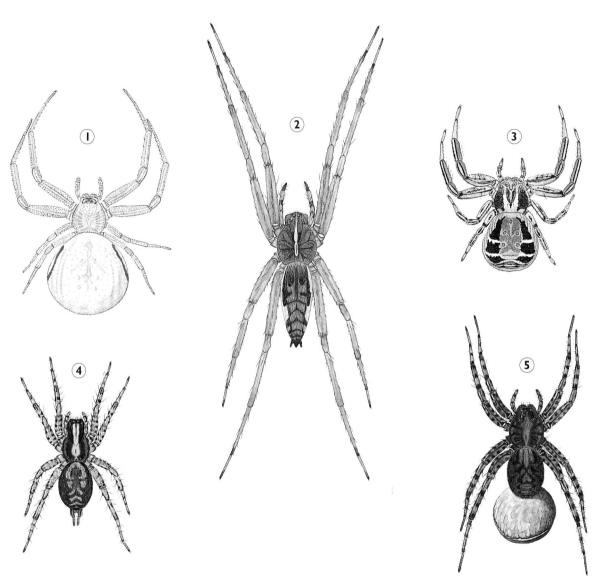

1 Common crab spider
Misumena vatia
SIZE AND DESCRIPTION
3–11 mm long. Female is white, yellow or greenish, with reddish stripes on each side of the abdomen (although these may be absent). The male is much smaller, with dark brown stripes on the abdomen. The two front pairs of legs are dark brown.
HABITAT Flowery habitats. Widespread in northern Europe, but more common in the southern part of its range.
FOOD/HABITS Seen in the summer. Sits in white and yellow flowers waiting in ambush for prey species.

2 Nursery-web spider
Pisaura mirabilis
SIZE AND DESCRIPTION
10–15 mm long. Sexes are similar, but males are smaller, with narrower abdomens. Colours vary from yellow to brown, with markings that may be very clear or even absent.
HABITAT Widespread in grassland, heathland, woodland and gardens across northern Europe.
FOOD/HABITS Seen in summer. Diurnal hunter. Runs swiftly and suns itself on plants. Female carries her egg-cocoon with her fangs. She later spins a silken tent over it and then stands guard until the young disperse.

3 Crab spider
Xysticus cristatus
SIZE AND DESCRIPTION
3–8 mm long. Females may be almost twice the size of males. Abdomen has triangular markings. Patterns are variable. Rather crab-like.
HABITAT Widespread throughout northern Europe in bushes, low plants and on the ground.
FOOD/HABITS Seen spring and summer. Hunts by lying in wait on flowers and pouncing on insect prey. Well camouflaged.

4 Toothed weaver
Textrix denticulata
SIZE AND DESCRIPTION
6–7 mm long. Males and females are similar in size and appearance. The spinnerets, located at the base of the abdomen, are especially prominent.
HABITAT Widespread throughout northern Europe, both in open countryside and homes.
FOOD/HABITS Females seen throughout the year; males in the summer only. This spider is often seen running over warm ground in summer. The web is a triangular sheet with a tunnel-like retreat at the apex.

5 Spotted wolf spider
Pardosa amentata
SIZE AND DESCRIPTION
5.5–8 mm long. Dark, variably patterned. Male is smaller than the female.
HABITAT Widespread throughout northern Europe where there is low-growing vegetation and open ground.
FOOD/HABITS Females are seen spring to autumn; males disappear after mid-summer. The female carries her eggs in a silken sac attached to her spinnerets. When young spiders hatch, they climb onto her back and are carried for a short time.

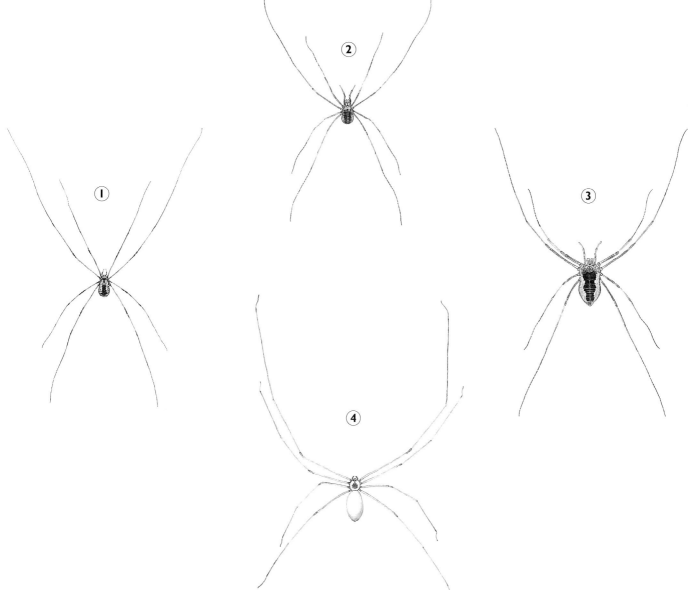

1 *Leiobunum rotundum*
SIZE AND DESCRIPTION
3–7 mm long. The female is almost twice as large as the male and has an oval, pale brown body with an almost rectangular patch on the back. The male has rusty-brown circular body with a black eye-turret on the back. Legs are long, very thin and almost black.
HABITAT Dense vegetation across most of Europe, except in the north and the south.
FOOD/HABITS Feeds on small invertebrates. Often seen resting by day on walls and tree-trunks.

2 *Opilio parietinus*
SIZE AND DESCRIPTION
5–9 mm long. Brown or greyish, with darker bands and sometimes a pale stripe running down the back. Females may have a saddle-like mark on the back. The underside is off-white with brown spots. Long, hair-like brown legs.
HABITAT Found on tree-trunks, bushes and rough grasses across Europe, except in the far north. Especially common around human habitation.
FOOD/HABITS Feeds on small invertebrates. Often seen on walls and fences.

3 Harvestman
Phalangium opilio
SIZE AND DESCRIPTION
5–8 mm long. The female is slightly larger than the male. Greyish or yellowish in colour, with a pure white underside. Very long, thin legs.
HABITAT Anywhere in Europe with dense vegetation.
FOOD/HABITS A nocturnal feeder on other small invertebrates. Winters as an egg and matures in late summer.

4 Daddy-long-legs spider
Pholcus phalangioides
SIZE AND DESCRIPTION
7–10 mm long. The daddy-long-legs has a cylindrical abdomen and very long legs. Colour is pale yellowish-grey.
HABITAT Rooms and cellars in buildings and also in caves. Occurs throughout central and southern Europe.
FOOD/HABITS Females seen throughout the year; males only in spring and summer. Hangs upside-down from a flimsy web, in which it catches flies and other spiders. Victims are trapped by having thread spun over them. When disturbed, the spider vibrates rapidly and spins to confuse predators.

Amphibians and reptiles

Amphibians are the most primitive class of land-living vertebrates. The young develop underwater and breath through gills before metamorphosing into adults that breathe through their skins or lungs. Although the adults leave water for much of the year, they tend to stay in damp places, because they must keep their skins moist and will die very quickly if they become dehydrated.

In Europe there are 43 species of amphibian belonging to two orders. The Urodela covers newts and salamanders and the Anura covers frogs and toads. Newts and salamanders are long bodied with strong tails and soft, moist, scaleless skins. All but three of the European species belong to one family. They breed in water, laying eggs that are deposited on plants or stones. The tadpoles develop their front legs first, but soon look like smaller editions of the adults, but not in colour and they have external gills. Like the adults the larvae are carnivores. Adult frogs and toads lack the tails of newts and salamanders and are anatomically adapted to springing. The females spawn in water and their eggs are fertilised by the male as they are laid. Common frogs lay the familiar frogspawn in masses of jelly. Common toads lay their eggs in a ribbon of jelly and the midwife toad tangles the ribbons of eggs around the male's legs. Frogs and toads are most commonly seen as adults away from water, while the newts and salamanders tend to be nocturnal and are not often seen during the periods of the year that they are not in water.

Most reptiles, which are covered in scales, have adapted to a terrestrial life and have a different reproductive cycle from amphibians. The eggs are fertilised internally and the young are miniatures of the adults. In most species young hatch from eggs, but a few species, such as the common or viviparous lizard, give birth to live young. Lizards, which possess four legs (which may not be visible, as is the case with slow worms), and snakes, which are legless, belong to the same order. Neither lizards nor snakes can maintain their body temperature internally, in the way that birds and mammals do. They have to maintain a more or less constant temperature by sunbathing and then moving into cover. They do have the advantage that little internal heat needs to be produced and they can therefore exist for long periods on very little food.

Amphibians and reptiles cannot survive very cold conditions and have to hibernate in winter in colder areas. In the far north of Europe the hibernation may last for as much as two-thirds of the year, while in southern Europe they may not need to hibernate at all. They seek a hole in the ground or crevice in which to hibernate. Frogs may hibernate underwater in the mud of ponds and ditches and toads have been found hibernating in old birds' nests.

The number of species of amphibians and reptiles diminishes in a northwards direction. Britain has fewer species than northern France. This is because as the last Ice Age retreated northwards about 10,000 years ago, amphibians began to move northwards. However, few species had reached what is now England before Britain became cut off from Europe as sea-levels rose and the English Channel and North Sea were created. This is why Ireland has even fewer species of amphibians and reptiles: the absence of snakes there, has more to do with rising sea levels than with St Patrick!

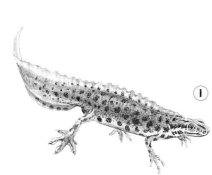

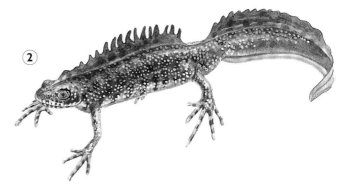

1 Common newt
Triturus vulgaris
SIZE AND DESCRIPTION Combined body and tail length: 7–11 cm.
Breeding males develop a wavy crest along the neck, back and tail.
Bright orange belly has black spots. Smaller females are less clearly
marked, lack the crest and have paler bellies.
HABITAT Damp places in a range of habitats. Across Europe from Ireland
to Italy. Not in southern Europe and northern Scandinavia.
FOOD/HABITS Eats insects, caterpillars, crustaceans, molluscs, worms,
tadpoles and slugs. Adults enter the water in February or March, leaving
it in June or July to hibernate in October.
BREEDING Pairs perform complex displays in the water. The male drops
his spermatophore, which is taken up by the female. Eggs are laid singly
on the leaves of water plants. Tadpoles emerge after two or three
weeks. They metamorphose in ten weeks, leaving as newts in August or
September. Sexually mature at three years. Lifespan is up to 20 years.

2 Great crested newt
Triturus cristatus
SIZE AND DESCRIPTION Combined body and tail length: 11–16 cm. Large,
colourful and warty. Upper part is dark brown or slaty-black. Undersides
are bright orange-yellow, spotted with black. Breeding males develop a
ragged crest along the back and another on the tail.
HABITAT Breeds in lowland water bodies, such as clay pits, reservoirs,
ditches and ponds, preferring pools 30–100 cm deep. Throughout
Europe, except Ireland, Iberia and northern Scandinavia.
FOOD/HABITS Hunts invertebrates and frog tadpoles at night. Enters the
water in mid-March, and remains until July or August. Hibernation begins
in October.
BREEDING Eggs are laid on leaves. Larvae metamorphose in four months.
Sexually mature at three years. Lifespan is up to 27 years.

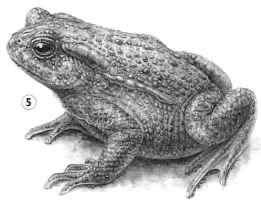

3 Common frog
Rana temporaria
SIZE AND DESCRIPTION 6–8 cm long.
Smoother skin and longer hindlegs than
common toad. Hindlegs are short compared
with other frogs. Colour and pattern vary.
Snout is rounded and the large black eyes
are surrounded by gold, flecked with brown.
Moves with a springing leap.
HABITAT Widespread in moist, shady
habitats. From northern Spain to the North
Cape. Absent from Iceland, Orkney and
Shetland.
FOOD/HABITS Snails, slugs, worms,
woodlice, beetles and flies are flicked
into the wide mouth by the long tongue.
Hibernates in pond mud or rotting
vegetation on land.
BREEDING Up to 1,400 eggs are laid.
Tadpoles metamorphose into froglets in 12
weeks, but stay near water until hibernating
in October or November. Sexually mature
at three years.

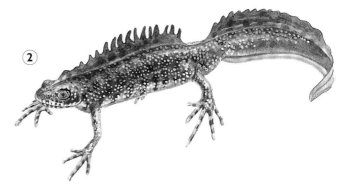

4 Midwife toad
Alytes obstetricans
SIZE AND DESCRIPTION Up to 5 cm long. Skin
is covered by small warts. Plump body varies
from ash to olive-brown. Eyes have vertical
irises. Moves by walking and springing. Bell-
like, trisyllabic call sounds like a scops owl.
HABITAT Gardens, fields, woods and scree
up to 2,000 m. South-west from Holland to
southern Spain. Introduced to England.
FOOD/HABITS Hunts insects, spiders and
worms at night.
BREEDING Female lays eggs two to four times
from March to August. Male wraps a string
of 20–100 fertilized eggs around his legs. He
carries them around for up to eight weeks
until they hatch.

5 Common toad
Bufo bufo
SIZE AND DESCRIPTION 8–15 cm long. Females
larger than males. Skin is warty and usually
orange-brown or olive. Walks and hops.
HABITAT Lives in a wide range of habitats,
but is usually found in damp places. Absent
from Ireland, northern Scandinavia and
Mediterranean islands.
FOOD/HABITS Insects, larvae, spiders, worms
and slugs are grabbed by the long, sticky,
prehensile tongue. Emerges from hibernation
to enter the water in February or March.
Hibernates again in October.
BREEDING Male clings to the female's back
and fertilizes the ribbons of 600–4,000 eggs as
she releases them. Tadpoles hatch in ten days
and metamorphose in two to three months.
Toadlets leave the pond in late July and
August. Lifespan is up to 40 years.

Reptiles

1 Grass snake
Natrix natrix
SIZE AND DESCRIPTION 70–150 cm long. Females are bigger than males. A slender, pale-coloured snake with a distinct head and dark marks on either side of the neck. Mouth looks curved. The viperine snake of France and Spain is very similar.
HABITAT Lowland hedgerows, woodland margins, heaths, moorland, water-meadows, gravel pits and gardens. England and Wales across Europe, north from Spain to southern Scandinavia and Russia.
FOOD/HABITS Eats frogs, fish, tadpoles, newts, fish, mice, voles and birds. Swims well. Hibernates from October to March in holes, crevices and manure heaps.
BREEDING Mating is in April and May; 8–40 eggs are laid from June to early August in manure heaps, haystacks, compost heaps and rotting logs. Eggs hatch in August or September. The young look like miniature adults. Males are sexually mature at three years, females at four years. Lifespan is up to 25 years.

①

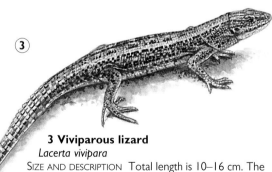

②

2 Slow-worm
Anguis fragilis
SIZE AND DESCRIPTION 30–52 cm long. A legless lizard with a round head and a smooth-scaled body. Brownish, but blue may show on older slow-worms.
HABITAT Meadows, woodland margins, gardens and cemeteries. From Britain and western Spain to Russia and southern Scandinavia.
FOOD/HABITS Hunts slow-moving invertebrates, usually early in the morning or in the evening.
BREEDING Mating is from April to June. The 6–12 young develop inside the female for three to five months before being born in August or September.

4 Common wall lizard
Podarcis muralis
SIZE AND DESCRIPTION Total length is 18–20. Tail can be more than twice the body length. Longer and more pointed head than the viviparous lizard. Colours vary from brownish or greyish to greenish. There are 14 European wall lizard species, so identification is difficult.
HABITAT Distributed from northern Spain across France and Italy to the Balkans and Greece. Found on walls and tree-trunks.
FOOD/HABITS Feeds on small invertebrates. Lives in colonies. Hibernates from November to February, except in warmer places.
BREEDING Mating is from February to June. Two to ten eggs are laid in a hole dug in the ground. They hatch from July to September.

③

3 Viviparous lizard
Lacerta vivipara
SIZE AND DESCRIPTION Total length is 10–16 cm. The tail may be twice the body length. Skin has obvious scales and a variable pattern. Females usually have a stripe down the middle of the back. Pale spots on the back are more obvious on the male. Pale underside may be orange in some males.
HABITAT Occurs across Europe to the Alps and northern Spain. In the south of its range, it lives in damp places up to about 3,000 m; in the north, it may be found in open areas such as overgrown and secluded gardens.
FOOD/HABITS Hunts by day using sight and scent. Prey includes spiders, insects and small snails. Hibernates from October to March.
BREEDING Young develop in thin membranous eggs inside the female's body. Eggs are laid from June to September, with the young "hatching" immediately.

④

Birds

The most familiar class of animal to be seen in gardens in Europe is birds. There are less than 10,000 species of birds in the world and about 420 species have been seen in Europe. Birds are distinguished from other classes of animals by their feathers, which have evolved from the scales of reptiles or dinosaurs. Their front limbs have evolved into wings and their bones are honeycombed with cells of air to provide the lightness necessary for flight. The breastbones of most birds are deep keels to which the strong muscles needed for flying are anchored. A bird's body temperature is normally maintained at 38–40°C, which is 3–4°C warmer than most mammals.

Most of the birds that visit gardens are species whose natural habitat is woodland glades and edges; habitats which closely resemble gardens. Many woodland species are members of the order Passeriformes, the perching birds and contain species with a highly developed capacity for learning, for example members of the crow family. However, if the garden has a pond with fish in it, it might be visited by birds, such as the grey heron and the much smaller kingfisher. Large gardens with large ponds might attract moorhens, which are members of the rail family.

The small perching birds that come so readily to bird-tables will also attract birds of prey. The sparrowhawk is a specialist hunter of small birds and the kestrel, which usually feeds on small mammals also takes birds in towns. The owls, which have adapted to nocturnal hunting have developed excellent vision. Their soft plumage enables them to hunt silently. Woodpeckers belong to the order Piciformes and have developed feet with two toes pointing forward and two pointing back, which enable them easily to climb up tree-trunks. They are also adapted to hacking at wood to find insects and grubs. Doves and pigeons are common garden visitors. Small-headed and thickly feathered, they have a distinct shape.

All birds lay eggs. Some do not build nests and their young emerge from the egg covered in down and are capable of moving on their own, and finding their own food. However, most birds in the garden make nests and their young hatch naked and completely dependent on their parents for food and warmth. The young grow fast and in some species of small birds the young become fully grown and feathered in little more than a fortnight of hatching. Some will rear more than one brood a year and one pair can rear a dozen young in a season. The mortality of small birds may be very high and the life expectancy of most small songbirds is about nine to twelve months from fledging. Predators, such as cats and sparrowhawks, will account for many bird fatalities. Others will die in accidents, such as collisions with windows or cars and many die from pesticides used in gardening, such as slug pellets.

Male

Female

Sparrowhawk
Accipiter nisus

SIZE AND DESCRIPTION 28–38 cm long, with a 60–75 cm wingspan. Wings are blunt and broad. Males, which are much smaller than females, have blue-grey heads and backs, and breasts barred with rusty-red. Females have grey-brown barring on their breasts and a pale "eyebrow".

VOICE Makes a monotonous ringing call near the nest.

HABITAT Woodland, parks, gardens and hedgerows. Breeds across Europe, but British populations are mainly resident.

FOOD/HABITS Small birds are the main food of sparrowhawks, which hunt by ambushing their prey.

BREEDING April to June. Nests in trees, usually on a branch next to the trunk. The four or five eggs are incubated for 42 days by the female alone. Young fledge after about 32 days.

Kestrel
Falco tinnunculus

SIZE AND DESCRIPTION 33–39 cm long, with a 65–80 cm wingspan. Distinctive long tail and pointed wings. Males have grey heads, black-tipped grey tails and dark-flecked russet backs. Females and juveniles lack the grey head, have brown tails with narrow bars and have more dark flecks on their backs.

VOICE Kestrels are noisy at the nest-site, with their rasping "kee-kee-kee-kee" call.

HABITAT Farmland, moorland and other open areas. Breeds in cities and towns, and may be seen flying over gardens. Resident across Europe, although north- and east-European populations migrate during autumn.

FOOD/HABITS Hovers above grassland or perches on trees and pylons, ready to drop down on rodents in the grass. Also feeds on small birds, large insects and lizards.

BREEDING April to June. Nests in old crow's nests, holes in trees or rocks and on ledges on buildings. Lays four or five eggs, which are incubated mainly by the female for about 28 days. Young fledge after four or five weeks.

Female

Male

Female

Male

Pheasant
Phasianus colchicus

SIZE AND DESCRIPTION Male is 75–90 cm long including 35–45 cm tail; female 53–64 cm. Wingspan of 70–90 cm. Males have long, golden, barred tails, green heads and red wattles. Some may have white rings around their necks. Females have shorter tails and are buffish-brown.
VOICE Gives a loud, hoarse metallic call.
HABITAT Woodlands, farmland with hedges, big gardens and reedbeds. Introduced to Greece from Asia 2,000 years ago, and now spread across much of Europe.
FOOD/HABITS Feeds on seeds, fruit, nuts and roots.
BREEDING April to June. Males have more than one mate. Up to 15 eggs are laid in a hollow in the ground. The eggs are incubated by females for about 25 days. Hatchlings are covered in feathers and capable of feeding themselves. They fledge after about 15 days.

Juvenile

Adult

Grey heron
Ardea cinerea

SIZE AND DESCRIPTION 90–98 cm long, with a wingspan of 150–175 cm. With its broad wings, slow, deep wing-beats, and legs stretched out behind, the grey heron looks very large in flight. Adults in breeding plumage have black crests, white necks and yellow bills. Juveniles are greyer and have darker bills.
VOICE Call is a harsh "frank".
HABITAT All types of waterways and wetlands. Will enter gardens, often early in the morning, in search of fish. Widespread in Europe, except in the far north. Some birds move south during winter.
FOOD/HABITS Feeds on fish and other animals by waiting beside the water and striking with its strong bill.
BREEDING Eggs laid from February to April. Breeding colonies build large nests of twigs in trees. Both sexes incubate the three to five pale greenish-blue eggs for up to 28 days. The young hatch downy and helpless, and remain in the nest for up to 55 days.

Birds

Feral pigeon
Columba livia

SIZE AND DESCRIPTION 31–33 cm long. Wingspan 62–68 cm. Black wing-bars and a white rump, many feral pigeons resemble the rock doves from which they originate. But colours may vary from white to very dark grey, and some may be pale fawn.
VOICE A soft cooing.
HABITAT Sea cliffs, towns and villages.
FOOD/HABITS Seeds, grain and discarded human food.
BREEDING Breeds throughout the year, but mainly in spring. The two eggs are incubated by both sexes for up to 19 days. Young take 30–35 days to fledge. Two or three broods each year.

Juvenile

Adult

Woodpigeon
Columba palumbus

SIZE AND DESCRIPTION Measuring 40–42 cm long and with a wingspan of 75–80 cm, this is the largest of the European pigeons. Adults have white rings around the neck, and a white bar across each wing. The wings make a clattering sound on take-off and landing. The body is noticeably large in flight.
VOICE A soft, often repeated "coo-coo-coo-cu-coo".
HABITAT Woodland, farmland, parks and gardens. Found across Europe, but not Iceland and the far north. Populations in eastern Europe move south-west during autumn.
FOOD/HABITS Eats seeds, berries and beechmast. Feeds in flocks throughout the winter.
BREEDING Has been known to breed throughout the year, but its main breeding season is from April to June. Nest consists of a raft of twigs on a branch. The two eggs are incubated by both sexes for 17 days. Young fledge within 35 days. May have three broods or more each year.

Stock dove
Columba oenas

SIZE AND DESCRIPTION 32–34 cm long, with a wingspan of 63–69 cm. Smaller and less chunky than the wood pigeon, with a noticeable black trailing edge to the black-tipped wings. It lacks the wingbars and white rump of the feral pigeon.

VOICE A monotonously repeated "roo-roo-oo".

HABITAT Woods and farmland, parks and large gardens. Breeds across Europe, except Iceland, with eastern populations moving south-west in autumn.

FOOD/HABITS Feeds on seeds and grain, often in flocks with wood pigeons.

BREEDING March to June. Nests in holes in trees, cliffs and buildings. The two eggs are incubated by both sexes for up to 18 days. Young fledge after four weeks. Two or three broods each year.

Turtle dove
Streptopelia turtur

SIZE AND DESCRIPTION 26–28 cm long, with a wingspan of 47–53 cm. A slender, delicate looking dove with chestnut and chequered black upperparts and dark underwings. Sexes are alike. Juveniles are duller in colour and lacks the adult's neck patch.

VOICE A soft call of "toorr, toorr" or "turr, turr".

HABITAT Winters in sub-tropical Africa, moves north in April to May, departs again from early August to October. Is a regular visitor in Britain.

FOOD/HABITS Feeds on plant and grass seeds. A shy bird but will share grain with poultry in farmyards.

Juvenile

Adult

Collared dove
Streptopelia decaocto

SIZE AND DESCRIPTION 31–33 cm long, with a wingspan of 47–55 cm. Slimmer than other pigeons. The back is brown buff, while the head and underparts are pinkish-brown. There is a black ring around the neck, and the wings have grey rumps and patches.

VOICE A rapidly repeated "koo-koo, koo" call.

HABITAT Towns, gardens and farmland with hedges. Has spread across Europe from Asia.

FOOD/HABITS Feeds on seeds and grain. A frequent bird-table visitor. Large flocks assemble at grain stores.

BREEDING March to September. Nest is a platform of twigs on a tree branch. The two eggs are incubated by both sexes. Young fledge within 18 days. Up to five broods may be reared each year.

Birds

Juvenile

Adult

Moorhen
Gallinula chloropus

SIZE AND DESCRIPTION 32–35 cm long. Wingspan 50–55 cm.
With its slaty plumage, very dark brown wings, white undertail
coverts, yellow-tipped red bill and green legs, the moorhen
is unmistakable. It flicks its tail as it walks with a careful tread.
When swimming, its head jerks forward and its tail points
upwards, giving the body a triangular appearance.
VOICE The varied repertoire includes harsh metallic "krrek" and
"kittick" calls.
HABITAT Ponds, rivers, canals, lakes and marshes across Europe.
Also found in parks and gardens with large ponds.
FOOD/HABITS Moorhens are omnivores, feeding on seeds,
insects, molluscs, leaves and carrion.
BREEDING Breeds from March to August. A cup-shaped nest
of plant material is built near the water and in low branches.
The clutch of 5–11 eggs is incubated by both sexes. Young are
feathered on hatching, but are not capable of feeding themselves.
They fledge in six or seven weeks. Two or three broods are
reared each year. Birds from earlier broods help to rear the
young of later broods.

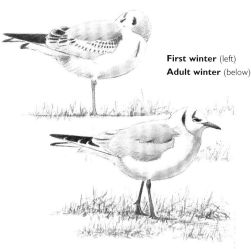

First winter (left)
Adult winter (below)

First summer (left)
Adult summer (below)

Black-headed gull
Larus ridibundus

SIZE AND DESCRIPTION 35–39 cm, with a wingspan of 86–99 cm. Most likely to be seen
in winter, when the head is white with a grey-brown crescent behind the eye. Immature
birds, whose backs and heads are mottled with cream and brown, moult into adult
plumage over two years. In breeding plumage, both sexes develop a chocolate-brown
head. Note the bill, which is red and finer than other European gulls'.
VOICE Noisy when in flocks. Calls include a strident "kee-yah".
HABITAT Breeds in colonies on moorland bogs, reedbeds, freshwater marshes and
lakes across northern Europe. In winter, it is common on ploughed fields, town parks,
playing fields, large gardens and coasts. Russian and east-European populations migrate
south-west during winter.
FOOD/HABITS Feeds on invertebrates, seeds and scavenges among rubbish.
BREEDING April to July. Nests in a scrape or mound of vegetation. Lays three buff eggs,
which are spotted with black. Both sexes incubate for up to 27 days. Young emerge
covered in down, and are fully fledged within six weeks.

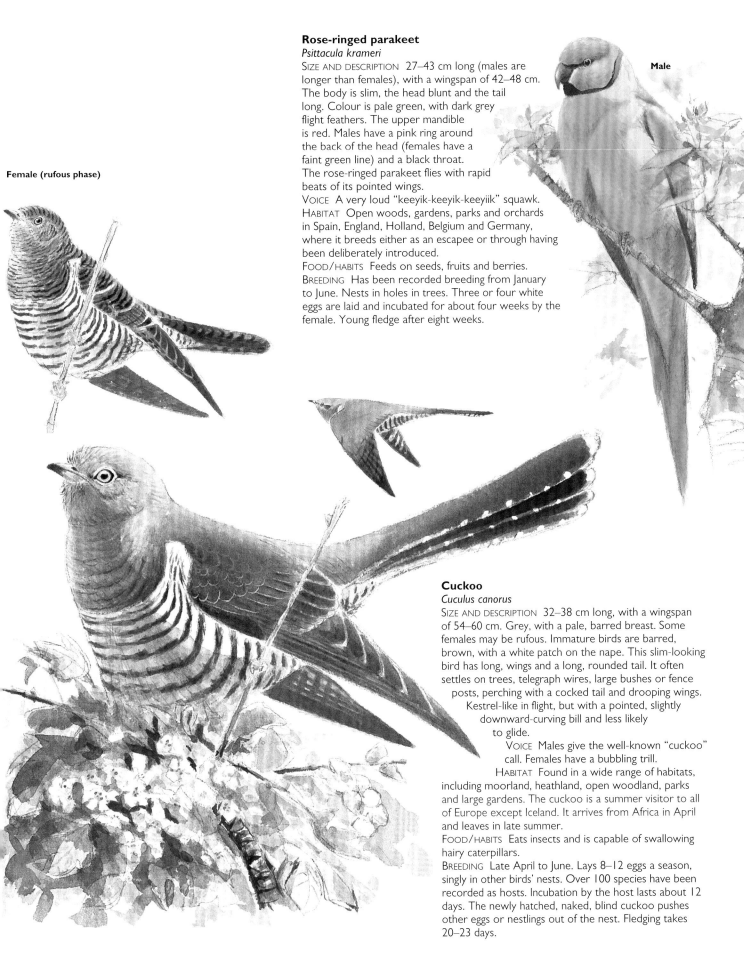

Rose-ringed parakeet
Psittacula krameri
SIZE AND DESCRIPTION 27–43 cm long (males are
longer than females), with a wingspan of 42–48 cm.
The body is slim, the head blunt and the tail
long. Colour is pale green, with dark grey
flight feathers. The upper mandible
is red. Males have a pink ring around
the back of the head (females have a
faint green line) and a black throat.
The rose-ringed parakeet flies with rapid
beats of its pointed wings.
VOICE A very loud "keeyik-keeyik-keeyiik" squawk.
HABITAT Open woods, gardens, parks and orchards
in Spain, England, Holland, Belgium and Germany,
where it breeds either as an escapee or through having
been deliberately introduced.
FOOD/HABITS Feeds on seeds, fruits and berries.
BREEDING Has been recorded breeding from January
to June. Nests in holes in trees. Three or four white
eggs are laid and incubated for about four weeks by the
female. Young fledge after eight weeks.

Male

Female (rufous phase)

Cuckoo
Cuculus canorus
SIZE AND DESCRIPTION 32–38 cm long, with a wingspan
of 54–60 cm. Grey, with a pale, barred breast. Some
females may be rufous. Immature birds are barred,
brown, with a white patch on the nape. This slim-looking
bird has long, wings and a long, rounded tail. It often
settles on trees, telegraph wires, large bushes or fence
posts, perching with a cocked tail and drooping wings.
Kestrel-like in flight, but with a pointed, slightly
downward-curving bill and less likely
to glide.
VOICE Males give the well-known "cuckoo"
call. Females have a bubbling trill.
HABITAT Found in a wide range of habitats,
including moorland, heathland, open woodland, parks
and large gardens. The cuckoo is a summer visitor to all
of Europe except Iceland. It arrives from Africa in April
and leaves in late summer.
FOOD/HABITS Eats insects and is capable of swallowing
hairy caterpillars.
BREEDING Late April to June. Lays 8–12 eggs a season,
singly in other birds' nests. Over 100 species have been
recorded as hosts. Incubation by the host lasts about 12
days. The newly hatched, naked, blind cuckoo pushes
other eggs or nestlings out of the nest. Fledging takes
20–23 days.

Hoopoe
Upupa epops

SIZE AND DESCRIPTION 25–29 cm long, with a wingspan of 44–48 cm. Upper parts are sandy-fawn, with black-and-white barred wings and tail, and the bill curves downwards. The hoopoe's remarkable crest is raised momentarily when the bird lands. The wings are broad and the flight is flappy, often low over the ground. It walks jerkily, with a starling-like gait. The hoopoe is surprisingly well-camouflaged when resting on the ground.

VOICE A repeated "poo-poo-poo".

HABITAT Gardens, vineyards, olive groves, and farmland with bushes. The hoopoe visits northern Europe between late April and September, moving south to North Africa and southern Spain in winter.

FOOD/HABITS Feeds on insects, worms and small reptiles found mainly on the ground or extracted from crevices with the curved bill.

BREEDING April to June. Nests in holes in trees, buildings or rocks. Five to eight eggs are laid and incubated for 16–19 days by the female. The helpless young fledge within four weeks.

Kingfisher
Alcedo atthis

SIZE AND DESCRIPTION 17–19.5 cm long. Wingspan 24–26 cm. The combination of bright blue and chestnut make the kingfisher unmistakable when perched above a pool, but these colours actually provide remarkable camouflage when the bird is resting among autumn leaves. The bill is black, but breeding females have a reddish base to the lower mandible. Juveniles have a pale spot at the tip of the bill.

VOICE A distinctive whistle.

HABITAT Rivers, streams and lakes. Will visit garden ponds to take small ornamental fish.

FOOD/HABITS Fish are the main food. The kingfisher hunts by diving into the water from a perch or by hovering and then diving.

BREEDING April to June. Nests in a chamber at the end of a tunnel, which can be between 45–90 cm long. Lays six or seven chalky-white eggs. Males and females share the incubation for up to three weeks. The helpless, naked young fledge within four weeks. May rear two broods each year.

Juveniles

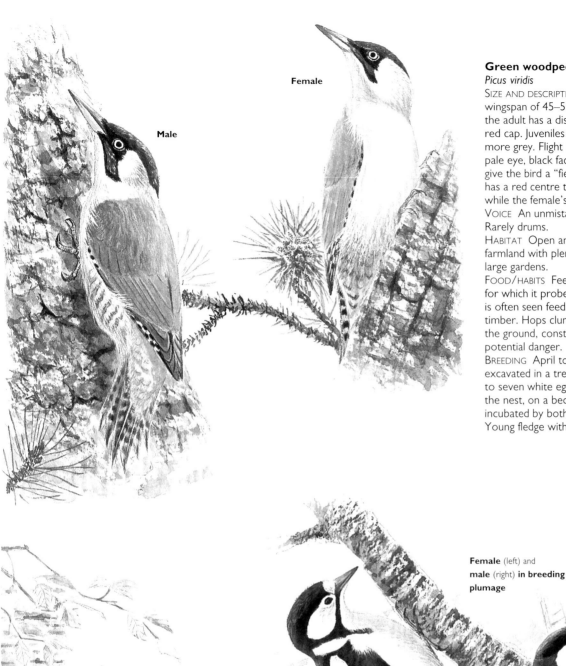

Male

Female

Juvenile

Female (left) and **male** (right) **in breeding plumage**

Green woodpecker
Picus viridis
SIZE AND DESCRIPTION 30–36 cm, with a wingspan of 45–51 cm. Green plumage, but the adult has a distinctive yellow rump and red cap. Juveniles are speckled, and appear more grey. Flight is deeply undulating. The pale eye, black face and moustachial stripe give the bird a "fierce" appearance. The male has a red centre to the moustachial stripe, while the female's is black.
VOICE An unmistakable shrill, laughing call. Rarely drums.
HABITAT Open and mixed woodlands, farmland with plenty of trees, parkland and large gardens.
FOOD/HABITS Feeds on insect grubs and ants, for which it probes rotten wood and soil. It is often seen feeding on lawns and decaying timber. Hops clumsily and nervously on the ground, constantly on the lookout for potential danger.
BREEDING April to June. Nest is a hole excavated in a tree-trunk by both sexes. Five to seven white eggs are laid on the floor of the nest, on a bed of wood chips, and are incubated by both sexes for 18–19 days. Young fledge within 21 days.

Great spotted woodpecker
Dendrocopos major
SIZE AND DESCRIPTION 23–26 cm long, with a wingspan of 38–44 cm. A blackbird-sized, black-and-white bird. It has white shoulder patches, with red under the tail. The male has a red patch on the nape, while the female's nape is black. Juvenile has a red crown. Flight is undulating. Similar species include the lesser spotted woodpecker, which is sparrow-sized.
VOICE A short, sharp "tchak" call, which may be repeated at 1-second intervals. In spring, it drums very fast on rotten branches.
HABITAT All kinds of woodland, large gardens and parks.
FOOD/HABITS Insects and grubs, and conifer seeds in winter. Will visit garden feeders. Also steals eggs and young from other birds' nests.
BREEDING April to June. Nests in holes excavated in tree-trunks. Four to seven white eggs. Incubation, mainly by the female, lasts 16 days. Young fledge in three weeks.

Swift

Apus apus

SIZE AND DESCRIPTION 17 cm long, with a wingspan of 40–44 cm. The swift has long, narrow, crescent-shaped wings, a torpedo-shaped body and a short forked tail. It has a dark brown plumage with a pale throat.

VOICE A shrill, monotone scream, which is often uttered by tight flocks flying round buildings at roof-top height.

HABITAT Breeds in towns and villages, but feeds in the sky, often several kilometres from nest-sites. A summer visitor to northern Europe (except Iceland), usually arriving in May and leaving in August.

FOOD/HABITS The swift is adapted to feed on high-flying insects, which it catches in its wide, gaping mouth. It has very short legs and shuffles around its nest-site. Most of its life is spent on the wing.

BREEDING May to July. Nests in holes in buildings in lose colonies. The nest is a rough pile of leaves and debris. Two white eggs are incubated for two or three weeks by both sexes. Fledging takes up to 48 days. Second broods are unusual.

Swallow

Hirundo rustica

SIZE AND DESCRIPTION 17–22 cm long, including a tail of 3–6.5 cm, juveniles are slightly smaller. The swallow's wings are long and pointed, and its tail deeply forked. It has pale cream underparts, dark blue wings and back, and a red throat with a blue-black neck band. Flight is fast, with powerful wingbeats.

VOICE In flight, it has a high-pitched "vit-vit" call. The warning call for cats (and other ground predators) is a sharp "sifflit"; for birds of prey it is a "flitt-flitt". The song is a rapid, rattling twitter.

HABITAT Breeds in farmyards and small-village gardens with surrounding open country. Often seen near water. It is a summer visitor to northern Europe, arriving in late March and April, and leaving in September and October. On migration, swallows roost in flocks in reedbeds and scrub.

FOOD/HABITS Feeds on insects, which it catches in flight by flying low over fields and water, manoeuvring to avoid obstacles.

BREEDING Late April to early August. Nests in outbuildings and porches, on rafters or shelves. Cup-nests are built from small pieces of mud, which are then lined with grass or straw. The four or five white eggs have reddish spots. The female (occasionally the male) incubates them for 14–16 days, and the young swallows fledge in 24 days. There are two, sometimes three, broods each year.

Adult

Juvenile

Pied/White wagtail
Motacilla alba

SIZE AND DESCRIPTION 17–19 cm long.
The male of the British race pied wagtail,
M. yarrelli, has a black back and wings,
and the female, a dark grey back. In the
continental race, the white wagtail, both
male and female have a pale grey back.
In flight, which is undulating, faint double
wing bars can be seen. On the ground,
the gait is rapid, with the head moving
back and forth and the tail wagging –
hence the name.
VOICE Typical flight call is a "chissick",
sometimes a "chissick-ick". Song is plain and
twittery.
HABITAT Towns, gardens and open habitat, usually near water. In
winter, although feeding singly or in small flocks, pied wagtails roost
in large flocks in warm places, such as around factories, glasshouses
and town centres. Scandinavian and east-European populations
migrate south-west in winter.
FOOD/HABITS Runs rapidly after flying insects,
sometimes snatching them with a small leap
into the air. Prefers feeding on lawns, roofs, car
parks and roads, where prey is easily spotted.
BREEDING April to June. Nests are untidily built
from fine twigs and grass, lined with hair and
feathers. Four to six white eggs, spotted with
grey. Incubation, mostly by the female, lasts
up to 14 days. Young fledge in 13–16 days.
Two broods each year.

Female Pied wagtail

Male White
wagtail

House martin
Delichon urbica

SIZE AND DESCRIPTION 12.5 cm long. Wings are broader than the
swallow's and the forked tail is shorter, giving the house martin a
stubbier appearance. The rump is white, while the wings, head and tail
are dark blue, appearing black in some lights. Flight is more fluttery than
the swallow's, and flaps are often interspersed with glides. Underparts
of juveniles tend to be a duskier white than on adults.
VOICE A harsh twitter, which becomes higher and more drawn-out
when agitated. Song is a series of formless chirps.
HABITAT Breeds in colonies in towns and villages, and sometimes
on cliffs. Arrives in Europe from Africa in March and April, leaving in
September and October.
FOOD/HABITS Tends to feed on flying insects at a
greater altitude than the swallow. Rarely seen on
the ground, except when collecting mud for
nest-building.
BREEDING May to early August. Buildings
provide a substitute for cliffs as nesting
sites for house martin colonies. Builds a
hemispherical nest under eaves with a small
entrance hole. Four to five white eggs are
incubated by both sexes for 19–25 days.
Two or three broods each year.

Juvenile

Adult

Wren

Troglodytes troglodytes

SIZE AND DESCRIPTION 9–10 cm long. The wren – a tiny brown bird with a short, upright tail – looks rather mouse-like on occasions. Its reddish-brown back is faintly barred, as are the paler flanks. There is a narrow dark eye-stripe, with a paler stripe above the eye. The bill is narrow, pointed and slightly downward-curving. The wren has a whirring flight, like a large bee.

VOICE Calls are a repeated "tic-tic" and a metallic "clink". The song is a surprisingly loud series of trills and warbles, delivered from prominent song-posts.

HABITAT Woodland with dense undergrowth, scrub, heathland, gardens, parks and moorland.

FOOD/HABITS Searches for insects and spiders on or near the ground, moving in a rather mouse-like way. In winter, flocks of wrens will roost together in a bundle on cold nights.

BREEDING April to July. Males build several nests in their territories. The female selects one, which is then lined with feathers. The nest is domed and well-camouflaged. The five to eight eggs are incubated by the female for 14–17 days. Young fledge in 15–20 days. Two broods each year.

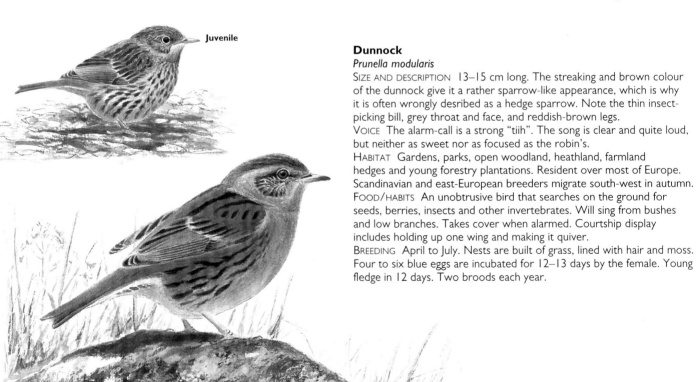

Juvenile

Dunnock

Prunella modularis

SIZE AND DESCRIPTION 13–15 cm long. The streaking and brown colour of the dunnock give it a rather sparrow-like appearance, which is why it is often wrongly desribed as a hedge sparrow. Note the thin insect-picking bill, grey throat and face, and reddish-brown legs.

VOICE The alarm-call is a strong "tiih". The song is clear and quite loud, but neither as sweet nor as focused as the robin's.

HABITAT Gardens, parks, open woodland, heathland, farmland hedges and young forestry plantations. Resident over most of Europe. Scandinavian and east-European breeders migrate south-west in autumn.

FOOD/HABITS An unobtrusive bird that searches on the ground for seeds, berries, insects and other invertebrates. Will sing from bushes and low branches. Takes cover when alarmed. Courtship display includes holding up one wing and making it quiver.

BREEDING April to July. Nests are built of grass, lined with hair and moss. Four to six blue eggs are incubated for 12–13 days by the female. Young fledge in 12 days. Two broods each year.

Juvenile

Female

Male

Blackbird
Turdus merula

SIZE AND DESCRIPTION 23.5–29 cm long. The all-black male, with its yellow bill and yellow eye-ring, is unmistakable because it is larger and has a comparatively longer tail than the starling. The sooty-brown female, with a dark-streaked pale throat, and the gingery juveniles may be confused with other thrushes, but they have a solid build and cock their tails when landing. First-winter males have all-dark bills.

VOICE Alarm call is a harsh "chack-aack-aack-aack", or a series of high metallic notes when going to roost or when a cat is seen. The song is a rich, melodic fluting, often rising to a crescendo.

HABITAT Woodland, parks, orchards and gardens. Blackbirds in eastern Europe and Scandinavia migrate in winter.

FOOD/HABITS Hops or walks over the ground, stopping and cocking its head to look for worms or other food. Takes a wide range of food, including insects, worms, fruit and berries.

BREEDING Early March to August. Nest is made of grasses, with a mud-cup lined with finer grasses. Lays four or five light blue, red-spotted eggs. Incubation by the female lasts 11–17 days. Young fledge in 12–19 days. Two or three broods each year.

Juvenile

Robin
Erithacus rubecula

SIZE AND DESCRIPTION 12.5–14 cm. Adopts a perky stance. The orange-red breast is fringed with pale grey, and orange covers the face. Underparts are pale and the back brown. The head seems rather large and the legs are comparatively long and thin. There is a pale wing-bar. Juvenile has a pale-spotted brown breast, and a pale-flecked head and back, with the wing-bar quite noticeable.

VOICE The call is a short, hard repeated "tic", repeated most rapidly when anxious and going to roost. The alarm call is a thin, sharp "tsiih". The song is sweet and starts high, followed by a fall, and then speeds up in clear, squeaky notes.

HABITAT The robin is a woodland bird that breeds in gardens, parks and forest edges. In winter, northern European robins migrate south-west to southern Europe. Other populations are resident.

FOOD/HABITS Feeds on berries and insects on the ground. Moves over the ground by hopping vigorously. In winter, the robin will search for food in mole-hills, animal tracks in the snow and where soil is being turned over by gardeners.

BREEDING March to July. Builds a cup-nest in tree-stumps, on branches, among ivy, and in open-fronted nestboxes. Five or six red-speckled white eggs are laid and incubated for 12–15 days by the female. Young fledge in 12–15 days. Two or three broods per season.

Song thrush

Turdus philomelos

SIZE AND DESCRIPTION 20–22 cm long. The song thrush has a brown back and a speckled creamy breast (speckles are shaped like arrowheads, and more regular than those of the mistle thrush). Shortish brown tail is not as short as the starling's. The underwing shows yellowish-orange. When the song thrush sings, the wing-tips can often be seen. Flies rather jerkily.

VOICE A loud, strong song, with a variety of trilling and squeaky notes with few pauses and frequent repetitions. The alarm call is a series of sharp, scolding notes, higher than the blackbird's. Contact call in flight is a fine, sharp "zit".

HABITAT Woodlands, parks and gardens. Populations from Scandinavia and eastern Europe winter in western and southern Europe. British populations have seriously declined in recent years.

FOOD/HABITS Feeds on worms, insects, berries and snails, the shells of which it smashes on hard ground or rocks. May be seen feeding on playing fields and in parkland in small flocks.

BREEDING March to August. Builds a neat cup-nest of grasses and fine twigs, lined with mud. Four to six pale blue eggs are incubated by the female for 11–15 days. Young fledge within 16 days. Two or three broods each season.

Redwing

Turdus iliacus

SIZE AND DESCRIPTION 19–23 cm long. The redwing is the size of a song thrush, but with a visibly larger head. The white stripe above the eye and the black-tipped yellow bill give it a more striking appearance than the song thrush. The red patch under the wing (redder than in the song thrush) is conspicuous when the bird flies. Flight is fast and direct.

VOICE On migration, redwings have a thin "tseep" contact call. The alarm call is hoarse, rattling and scolding. Song is variable, with loud fluted notes followed by a prolonged twitter.

HABITAT Breeds in upland birch and conifer forests in Scandinavia and north-eastern Europe. It migrates south-west in winter to western and southern Europe, where it feeds in hedges, moving to open fields and gardens as hedgerow food runs out.

FOOD/HABITS Feeds on worms, insects and berries. Redwings and other thrushes are attracted to gardens by berry-bearing shrubs, such as cotoneaster and pyracantha.

BREEDING May to July. Builds a cup-nest against a tree-runk. Four or five pale blue, brown-speckled eggs are incubated by the female for 11–15 days. Young fledge in 11–15 days. Two broods per season.

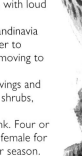

Fieldfare
Turdus pilaris

SIZE AND DESCRIPTION 22–27cm long. Smaller than the mistle thrush, but with a stockier appearance. The grey head, red-brown back and the apricot background to the speckled breast make this a striking bird. The speckles become downward facing arrowheads. The longish tail and pale grey rump show clearly in flight, as do the white underwings. Flight is more flapping and less undulating than the mistle thrush's. The dark face-markings and white eyebrow give the fieldfare rather a fierce expression.

VOICE The most frequently heard call is a harsh "chack-chack-chack". On migration, the flight call is a thin "see". The song is a tuneless, chattering babble.

HABITAT Breeds in Scandinavia and central and eastern Europe, in open forest with tall birches and pines and in town parks. In tundra, it uses buildings and other artificial structures. The fieldfare moves south in October and November to feed in fields and gardens when food becomes scarce.

FOOD/HABITS Feeds on worms, insects, berries and fruit. Particularly fond of windfall fruits in gardens and orchards during hard weather. Can be aggressive near the nest and will fly at crows and other corvids, sometimes bombing them with droppings.

BREEDING April to July. Builds a cup-nest high in a tree, often near the trunk, and nests in loose colonies. Five or six pale blue eggs with reddish markings are incubated by the female for 11–14 days. Young fledge in 11–14 days. One or two broods each season.

Mistle thrush
Turdus viscivorus

SIZE AND DESCRIPTION 22–27 cm long. This largish thrush has an upright and comparatively longer tail than the song thrush. Its white breast is speckled with rounded, blotchy spots. In flight, the white outer tail feathers and narrow white wing-bars can be seen. The underwing is white. On the ground, the mistle thrush stands in an upright posture. Flight is more undulating than the song thrush's.

VOICE Its flight call is a dry, churring rattle. Song is full-blooded, but similar in form to the song thrush's. It is often performed in bad weather, such as rain, when other birds are not singing.

HABITAT Breeds in open woodland, orchards, and parks and gardens with trees. Moves into fields and parkland to feed in winter.

FOOD/HABITS Eats worms, berries and insects. Less likely to be seen in flocks than most thrushes, but will feed alongside other species. May be seen in family flocks of about half a dozen birds. Those breeding in Scandinavia and eastern Europe move south in winter.

BREEDING March to July. Nests are built in tree-forks. Lays four or five blue eggs spotted with red. Eggs are incubated by the female for 12–15 days. The young fledge in 20 days. Two broods per season.

Birds

Chiffchaff

Phylloscopus collybita

SIZE AND DESCRIPTION 10–12 cm
long. A small, neat bird, with a fine
bill and thin legs. The similarity between
the chiffchaff and the willow warbler makes
it difficult to tell them apart in the field. To do so successfully, one
needs to hear the song, to spot behavioural differences or to have
very good views of the colour and the length of the wings. The legs
are usually dark and the bill is even finer than the willow warbler's. The
stripe above the eye is less distinct and shorter, while the darkish patch
beneath the eye emphasises the white eye-ring. The primary feathers do
not project very far beyond the tertial feathers. The chiffchaff's habit of
flicking its tail downwards is characteristic.
VOICE The call is a soft "hueet", and the song is a distinctively slow and
measured "chiff-chaff-chiff-chaff".
HABITAT Usually breeds in open deciduous woodland with some scrub.
Mainly a summer visitor to the British Isles, Scandinavia and central
Europe, arriving in mid-March and leaving from August to November.
Most likely to be seen in gardens while in transit.
FOOD/HABITS Feeds on small insects, which it finds by flitting around
among foliage. Eats berries in autumn.
BREEDING April to June. Builds a domed nest on the ground. Four to
nine white eggs, speckled with purple, are incubated by the female for
13–14 days. Young fledge in about 14 days. One or two broods.

Juvenile

Willow warbler

Phylloscopus trochilus

SIZE AND DESCRIPTION 11–12.5 cm long. The pale "eyebrow" (the
supercilium) is its most obvious feature. The head, back and tail are
generally brownish-green, and the throat and eyebrow are yellowish. In
northern Europe, willow warblers are more grey and less yellow. Young
birds are yellower. Legs are usually pale brown. Primary feathers project
well beyond the tertials.
VOICE The call is a soft "huitt", similar to the chiffchaff's. The song is
rather sad and languorous, with clear notes descending and falling away.
HABITAT Breeds in upland birchwoods and other deciduous woods,
as well as in parks and wooded gardens across Europe. A summer
visitor that winters in sub-Saharan Africa. Arrives March, leaves July to
September.
FOOD/HABITS Feeds on small insects found among leaves.
BREEDING April to June. Builds a domed nest on the ground. Lays six
or seven reddish-speckled white eggs, which are incubated for 13 days
by the female. Young fledge in 13–16 days. One or two broods.

Goldcrest
Regulus regulus

SIZE AND DESCRIPTION 8.5–9.5 cm long. This tiny bird has a greenish back, and a yellow crest that becomes orange in the male. The crest has a black stripe on each side. The face is greyish, with dark eyes surrounded by very pale grey.

VOICE A very high-pitched, thin call of three or four syllables: "see-see-see". Song is high-pitched and rhythmic, and ends with a trill or flourish.

HABITAT Coniferous and mixed woodlands are the favoured breeding habitat, with spruce and fir preferred. In gardens, goldcrests are often seen in yew and cypress trees. Birds from the north of Scandinavia move south in winter.

FOOD/HABITS Tiny insects and spiders are the food of the goldcrest, which feeds on the undersides of leaves and branches, sometimes hovering at the tips of branches to pick off insects. In winter, it will join flocks of tits foraging in woodland.

BREEDING May to early July. The nest is a cup of feathers and mosses, built high up in a tree, often near the tip of a branch. Lays seven to ten brown-speckled white eggs, which the female incubates for about a fortnight. Young fledge within three weeks. Two broods per season.

Female

Male

Spotted flycatcher
Muscicapa striata

SIZE AND DESCRIPTION 13.5–15 cm long. With a greyish brown back and pale underparts, the spotted flycatcher is not a striking bird, but close examination reveals its streaked forehead and faintly streaked upper breast. Its bill and legs are black, and its black eye is an obvious feature. When perched, its posture is upright.

VOICE The call is a short, shrill "tzee". The song is quiet, simple and scratchy, often with soft trills.

HABITAT Open woodland, parks and gardens are the preferred habitat of this summer visitor, which arrives at the end of April or beginning of May and leaves in September.

FOOD/HABITS The hunting flight of the spotted flycatcher is characteristic. The bird takes off from a perch, snatches a flying insect in flight and then returns to the same perch. It will hover briefly when taking larger insects, such as butterflies and damselflies.

BREEDING May and June. Builds a cup-nest between a branch and tree-trunk, or among espaliered trees against the wall. Also nests in open-fronted nestboxes. Four or five pale blue eggs are incubated by the female for 11–15 days. Young fledge in a fortnight. One or two broods per season.

Juvenile

Birds

Blue tit
Parus caeruleus

SIZE AND DESCRIPTION 11–12 cm long. Smaller than the great tit and possessing a bright blue crown. The stripe down the yellow breast is less well-defined than the great tit's. The tail and wings are blue. Young birds have yellow cheeks, and the blue parts are green.

VOICE A clear, ringing, high-pitched song, and a thin "see-see" call.

HABITAT Mixed and deciduous woodlands, parks and gardens. Found across Europe, except in Iceland and northern Norway.

FOOD/HABITS Feeds on insects, spiders and other small animals, finding them on tree branches and sometimes in the corners of windows. Often visits the bird table in winter. Feeds in flocks of up to 30 in winter, often with other species of tit.

BREEDING April to May. Builds a cup-nest in a hole in a tree or a nestbox. Lays 7–12 eggs which are incubated by the female for up to 16 days. Young fledge in about three weeks.

Juveniles

Juvenile

Great tit
Parus major

SIZE AND DESCRIPTION 14 cm long. A black cap and a black stripe starting at the bill give the great tit a more ferocious expression than the blue tit. The male's breast-stripe becomes broader than the female's, and his colours tend to be more intense. Young birds have yellow cheeks for a few weeks.

VOICE The great tit's rich and varied repertoire includes a metallic "pink" and a repeated "teacher-teacher".

HABITAT Woodlands and gardens. Many of the tits feeding in gardens in winter return to woods to breed in spring. Found across Europe, except in Iceland and northern Norway.

FOOD/HABITS Feeds on seeds and fruits. Also takes spiders and insect larvae, such as sawfly caterpillars, in the breeding season. Eats sunflower seeds, peanuts, and fat at bird-tables. Feeds on the ground and in trees.

BREEDING March to May. Nest-sites are holes in trees, but it also uses nestboxes. Builds a cup nest in which 8–13 eggs are laid. Incubation by the female takes a fortnight and the young fledge in about three weeks.

Long-tailed tit
Aegithalos caudatus
SIZE AND DESCRIPTION 12–14 cm long, including a tail that is at least as long as the dumpy body. With its pink, black and white body and long tail, the long-tailed tit is unmistakable.
VOICE A piercing, trisyllabic continuous call "zee-zee-zee".
HABITAT Woods with bushy undergrowth, hedges and gardens.
FOOD/HABITS Feeds mainly on insects and small spiders, but is increasingly visiting bird-tables. Families form into flocks and move through woods and hedges, often with other tits.
BREEDING March to May. Constructs an exquisite, dome-shaped nest of moss and feathers, in which up to 12 eggs are laid. Incubation is mainly by the female. The young fledge within 18 days. One or two broods per year.

Coal tit
Parus ater
SIZE AND DESCRIPTION 11.5 cm long. Smaller than the great tit, with a proportionately larger head. The black head has white checks, and there is a white patch on the nape. The back is grey and the breast is grey-brown. The double wing-bar shows in flight.
VOICE The most frequent call is a triple "tsee-tsee-tsee". The song is like a simpler, weaker great tit's song.
HABITAT Woodlands and gardens. Prefers coniferous trees. Found across Europe, except in Iceland and northern Norway.
FOOD/HABITS Eats insects and seeds, particularly spruce cones in the north.
BREEDING April to June. Nests in holes in trees. Seven to nine eggs are laid and incubated for about a fortnight by the female. Young fledge in about 19 days. One or two broods per season.

Juvenile

Nuthatch
Sitta europaea

SIZE AND DESCRIPTION 12–14.5 cm. The large head, lack of a neck, short tail and heavy pointed bill give this sparrow-sized bird a distinctive shape. The back and head are slate-grey, with a long black eye-stripe. The cheeks are white and the breast and underparts are rusty-orange (darker in the male than in the female). In Scandinavian nuthatches, males have white breasts and the females have pale orange. Flight is undulating like a woodpecker's, but the tail-shape is rounded.

VOICE The nuthatch reveals its presence with a loud, strident "hwitt" call . The song is a repetitive "peeu-peeu-peeu".

HABITAT Breeds in mixed and deciduous woods, and in parks and gardens where there are mature oaks. It is resident from western Russia across Europe, but absent from Ireland, Scotland and all but the southernmost part of Scandinavia.

FOOD/HABITS Feeds on nuts, seeds and invertebrates. It uses its pointed bill to winkle spiders and insects out of bark crevices, and also to wedge nuts into crevices so it can then hammer holes into them. Nuthatches often descend tree-trunks head-first.

BREEDING April to May. Breeds in holes in trees and nestboxes. The nest entrance is plastered with mud to make it smaller. Lays six to nine white eggs, which have reddish patches. Incubation by the female takes up to 18 days. Young fledge in 23–25 days. A single brood per season.

Treecreeper
Certhia familiaris

SIZE AND DESCRIPTION 12.5–14 cm long. Mottled brown on the back and white on the underside camouflage this small bird as it climbs tree-trunks. The fine bill curves downwards. The long, stiff tail helps the bird balance in an upright position against a trunk. The short-toed treecreeper (absent from the British Isles) looks very similar, but has brown flanks, a different call and is found in more lowland habitats.

VOICE The call is a loud, thin "zzrreet". The song is several scratchy notes ending in a thin trill.

HABITAT In the British Isles and Scandinavia, the treecreeper is found in almost all woodland, and in parks, gardens and orchards with old trees with loose bark for nest-sites. In southern Europe, it prefers coniferous woodland and tends to occur at higher altitudes than the short-toed treecreeper. Both species are resident.

FOOD/HABITS The treecreeper probes bark crevices with its bill for insects and small spiders. It starts at the foot of a trunk and spirals upwards in a series of jerky movements. Then it flies down to the foot of the next tree and starts again. Cold winters can kill these birds. They roost in depressions in the soft bark of redwoods.

BREEDING April to June. The nest – a cup of twigs, lined with hair and feathers – is built behind flaking bark against a tree-trunk. Five or six reddish-speckled white eggs are incubated by the female for about a fortnight. The young fledge after about 14 days. Usually only one brood.

Juvenile

Starling
Sturnus vulgaris

SIZE AND DESCRIPTION 19–22 cm long. Superficially like a blackbird because of its black plumage and yellow bill, the starling differs in several key respects: it has a short tail and neck, an upright stance, pink legs, white spots and a metallic green shine. It also tends to be seen in flocks. In flight, the short tail and pointed wings give the bird an arrowhead profile. Flocks fly in tight formation, and the swirling flocks of thousands of birds at winter roosts can be quite dramatic. Juveniles are grey brown, and as they moult into adult plumage they may be seen with brown heads and speckled bodies.

VOICE The starling is a versatile mimic of other birds, but its own calls are creaky twitters, chirps, clicks and whistles. Its alarm call when a hawk is sighted is a sharp "kyett".

HABITAT Naturally a bird of oak woodland, the starling has spread into a number of habitats and seems particularly fond of human settlements. Populations over much of Europe are resident, but Scandinavian birds and those from east of Germany move south in winter.

FOOD/HABITS Starlings eat a wide variety of food. In winter, large flocks forage in fields and gardens, as well as on the seashore. Outside the breeding season, starlings form huge flocks that often roost in town centres.

BREEDING April to May. Nests in holes in buildings, trees and cliffs. Five or seven pale blue eggs are incubated by both sexes for 12–15 days. Young fledge in about three weeks. One or two broods each year.

Jackdaw
Corvus monedula

SIZE AND DESCRIPTION 30–34 cm long, with a wingspan of 64–73 cm. The smallest of the "black" crows, the jackdaw is not entirely black. Its nape is grey and its eye has a very pale iris. On the ground, its stance is more upright than either rook or carrion crow, and it struts as it walks. In flight, the wing-beats are faster and deeper than the crow's. The wings appear longer and narrower in proportion to the body, and the small bill makes the head look smaller. It flies in flocks almost as densely as pigeons.

VOICE Its most common calls are a metallic, high-pitched "kya" and "chak".

HABITAT Found throughout Europe on coasts, in ancient woodland and near human settlements.

FOOD/HABITS An omnivore, the jackdaw will eat invertebrates, such as worms and insect larvae, the eggs and nestlings of other birds, small mammals and grain. It will also forage on rubbish tips and may feed alongside other corvids in fields.

BREEDING April to May. Natural nest-sites are cavities in old trees or holes in cliffs and rock-faces. Buildings provide equally acceptable sites, and jackdaws are often found nesting in chimneys, churches and ruins. Four to six pale blue eggs are incubated by the female for 17–18 days. Young fledge in about four weeks. A single brood per season.

Birds

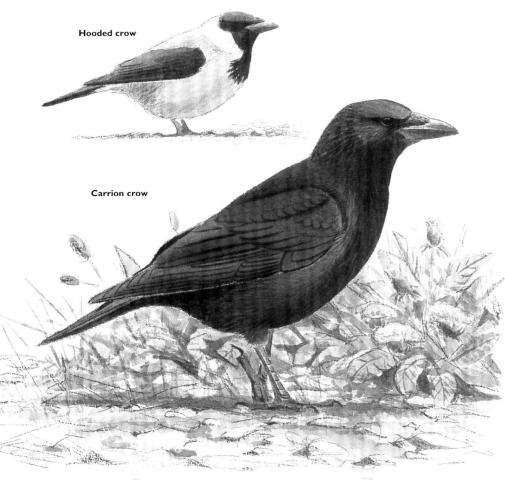

Hooded crow

Carrion crow

Carrion/Hooded crow
Corvus corone

SIZE AND DESCRIPTION 44–51 cm long, with a wingspan of 84–100 cm. The carrion crow is totally black, with a stout bill. The hooded crow, the subspecies found in eastern Europe, has grey underparts and a grey back. In flight, the wingbeats are shallow, the tail is rounded and the wings are uniformly broad.

VOICE A hard, rolling "krra-kra-kraa" is the commonest call.

HABITAT The carrion/hooded crow is found in a wide variety of habitats, from the coast to the mountains, including towns, parks and large gardens. The carrion crow is found in England, Wales and Scotland (except in the far north) and on the continent from Germany to Portugal, north of the Alps. The hooded crow is found in Ireland, the far north of Scotland and Europe east of Germany, and from the Alps south. In winter, hooded crows from north-eastern Europe move south-west to northern France.

FOOD/HABITS An omnivore, this crow feeds on carrion, nestlings and eggs, grain and insects. It is less sociable than the rook, but may still be seen in flocks.

BREEDING April to May. Builds a large cup-nest of twigs and sticks in trees, cliffs and on buildings. Lays four to six brown-speckled greenish-blue eggs, which are incubated for 18–20 days by the female. Young fledge in four or five weeks. A single-brood per season.

Magpie
Pica pica

SIZE AND DESCRIPTION 40–51 cm long, of which 20–30 cm is the tail. The overall impression of the magpie is of a black-and-white bird, but the wings are really a metallic blue-black and the long, round-tipped tail has a metallic green sheen. Males are larger and tend to have longer tails than the females. Flight is often a series of flaps interspersed with swooping glides. It walks with a strong gait, holding its tail well above the ground.

VOICE The magpie's noisy alarm call is a staccato, rather machine-gun-like rattle. Other magpie sounds include a variety of bisyllabic calls.

HABITAT Breeds around farms and villages and in hedgerows. It is becoming increasingly common in urban areas. Occurs throughout Europe, except in Iceland and northernmost Scotland.

FOOD/HABITS The magpie is an omnivore that feeds on seeds, insects, carrion (it is often seen feeding on roadside casualties), nestlings and eggs. It sometimes gathers in flocks of up to 25 birds.

BREEDING Starts nest-building early but may not lay eggs until March. Nests are football-sized twig domes built in trees and large hedge bushes. The five to eight pale blue eggs, blotched with olive, are incubated by the female for 17–18 days. Young fledge in 22–28 days. Only one brood per season.

Juvenile

Rook
Corvus frugilegus
SIZE AND DESCRIPTION 41–49 cm long, with a wingspan of 81–94 cm. The rook has a high forehead, purple-black plumage, and greyish skin around the base of the pointed bill. The juvenile does not have the greyish face until February or March after it hatches, and also has a black bill; it can be differentiated from the carrion crow by its high crown and waddling walk. The rook's feathery thighs give the appearance of bushy breeches. In flight, the trailing edges of the rook's wings narrow towards the body, giving a curved appearance, and the tail is almost wedge-shaped.
VOICE A hoarse, nasal croak.
HABITAT Feeds in open fields, but nests in woods and in tall trees in villages. Found in north-west Europe, including the British Isles and central and eastern Europe, but is very scarce in Scandinavia.
FOOD/HABITS Although it is omnivorous, it eats mainly insects and seeds, for which it forages in flocks in fields, sometimes with jackdaws.
BREEDING March to June. Large nests of twigs are built in colonies in woods or in mature trees in villages. From three to five brown-blotched blue-green eggs are incubated by the female for 16–20 days. Young take up to 30 days to fledge. One brood per season.

Jay
Garrulus glandarius
SIZE AND DESCRIPTION 32–35 cm long, with a wingspan of 54–58 cm. The pinkish-brown body contrasts with the white rump and black tail. The streaked feathers on the forehead are often raised in a crest. The jay has a pale eye, a black moustache and a blue-and-black wing-flash. In flight, the white rump, black tail and white wing-flash are very obvious. It flies with fluttery wingbeats and appears to make unsteady progress.
VOICE The call is a noisy, screeching "kscharch".
HABITAT Found in all types of woods, but prefers those with a plentiful supply of acorns. Has taken to gardens in some suburban areas. Found across Europe, where it is mostly resident, although in northernmost forests it may move south in winter if food is scarce.
FOOD/HABITS Feeds on seeds, fruits and the eggs and nestlings of other birds. Buries acorns, beechnuts and hornbeam seeds for later use.
BREEDING April to July. Builds a rather flat nest of twigs in a tree. The five to seven pale green eggs are speckled with buff, and are incubated by both sexes for 16–17 days. Young fledge in about 20 days. A single-brood only per season.

109

Birds

Male

Juvenile

Female

Bullfinch

Pyrrhula pyrrhula

SIZE AND DESCRIPTION 15.5–17.5 cm long. A compact, bull-necked finch, with a black cap and low forehead. The male has a rosy red breast, grey back, white rump and black tail. The female has a pale brown breast. Juveniles have grey-brown heads to match the breast-colour until September/October of their first autumn. Both sexes have white wing-bars that show clearly in flight. Flight is fast and gently undulating.

VOICE Call is a soft, rather melancholic fluted whistle.

HABITAT Breeds in mixed woodland, parks, large gardens and churchyards. Feeds in orchards, hedges and gardens.

FOOD/HABITS Buds and seeds are the main food, and the bullfinch's depredations on fruit trees and flowers can distress gardeners. Ash seeds are a favoured food, and some insects are taken in summer. Despite the colour of the male, the bullfinch is an unobtrusive bird that may easily be overlooked. It is usually seen in pairs or in small, loose flocks. It is resident in most of its range, although those breeding in northern Scandinavia and Russia move south in winter.

BREEDING Late April to July. The rather frail-looking nest of twigs supporting a cup of fine roots is built in a bush. Four or five blue eggs, spotted with purple, are incubated for a fortnight by the female. Young fledge in 12–18 days. One or two broods a year.

Chaffinch

Fringilla coelebs

SIZE AND DESCRIPTION 14–16 cm long. The bright colours of the male chaffinch in spring make it hard to confuse with other species. In winter, the blue-grey of the head and pink of the breast are subdued. The female is less easy to differentiate from other small brown birds, and it is most likely to be confused with the female house sparrow, although it is slimmer, has longer legs, a pale patch on the nape and no bold streaks on the back. Two white bars on each wing can be seen during the chaffinch's strong undulating flight.

VOICE The call is a sharp "pink", but the flight call is a softer "yupp". The song is a loud, ringing trill that becomes lower and then ends in a flourish, before being repeated again. Chaffinches demonstrate dialects in different districts.

HABITAT This very common bird breeds in all types of woodland, as well as in parks and gardens. Flocks form in autumn. Among the populations of Scandinavia and eastern Europe, which migrate south in winter, the flocks tend to be of single-sex. British chaffinches are resident, but birds from elsewhere in Europe may winter in Britain.

FOOD/HABITS Eats fruits and seeds, and also takes insects during the breeding season. Feeds on the ground, but will snatch flying insects in the air.

BREEDING The nest is a neat cup of mosses and lichens, lined with feathers and built in the forks of branches in small trees and bushes. Lays four or five pale blue eggs streaked with red. The eggs are incubated by the female for 11–13 days. Young fledge in 12–15 days. One or two broods a year.

Male

Female

Male

Juvenile

Female

Greenfinch
Carduelis chloris

SIZE AND DESCRIPTION 14–16 cm long. The greenfinch is stouter than most other finches. In summer, adults are olive-green, merging into grey-green on the face, wings and flanks, with bright yellow wing feathers on either side of the tail. The female's colouring is subdued, with faint brownish streaks on the back. The juvenile is paler and even more streaked. Females and juveniles could be mistaken in some lights for female house sparrows. Flight is bouncing and undulating.

VOICE Flight call is a sharp "burrurrup", while the song is a wheezy sequence of twitters and whistles.

HABITAT Breeding habitats comprise woodland edges, open woodland, parks, gardens and farmland with hedges. In winter, flocks may be seen feeding in farmland and gardens. Breeds across Europe, with northern Scandinavian breeders moving south in winter.

FOOD/HABITS Seeds and berries, along with some insects during the breeding season. A visitor to garden bird-tables, where it has acquired a taste for peanuts.

BREEDING Late April to early May. Four to six pale blue eggs, spotted with black, are incubated for up to a fortnight by the female. Young fledge in 13–16 days. Two or three broods per season.

Hawfinch
Coccothraustes coccothraustes

SIZE AND DESCRIPTION 16.5–18 cm long. This large finch has a big head, huge bill and short tail. The bill is blue-black in summer, becoming pale brown in winter. Sexes look very similar, but the male's flight feathers are completely black, while the female's secondaries are grey. In flight, broad white wing-bars and a broad white band on the tip of the tail are visible. Flight is undulating.

VOICE A sharp, robin-like "tic" from the tree canopy is probably a hawfinch. It also makes this call in flight. The song is a soft, stumbling series of "zih" and "zri" notes.

HABITAT Breeds in deciduous and mixed woodland with mature oaks, elms, ash, hornbeam and beech, and in old forest pasture with olives and cherries. In Scandinavia and Germany, it will visit garden bird-tables in winter. The hawfinch breeds across Europe, except in Ireland and Iceland. Many Scandinavian and east-European hawfinches move south in winter.

FOOD/HABITS Feeds on seeds, cherry-stones and nuts, splitting them with its powerful bill, which can exert a force of 50 kg. Some insects are also taken in the breeding season. Hawfinches are very elusive birds during the breeding season, but less shy in winter.

BREEDING April to July. A twiggy nest is built in tree branches. Five grey-blue, spotted eggs are incubated by the female, with occasional help from the male, for 9–14 days. Young fledge within 14 days. Single brood only.

Female

Male

Birds

Male

Female

Juvenile

Siskin
Carduelis spinus
SIZE AND DESCRIPTION 11–12.5 cm long. A small, neat bird, with dark-streaked, greenish-yellow plumage. The male is yellower than the female, and has a black cap and bib. Wing-bars in both sexes are yellow, and the male's tail has yellow patches on either side. Tails are deeply notched. Flight is rather flitting and uneven.
VOICE The flight call is either a descending "tilu" or a rising "tlui". Twittery, trilling song.
HABITAT Breeds in coniferous and mixed forests in Scandinavia and eastern Europe, as well as in parts of Britain, Ireland, Spain, France and Sardinia. Northernmost breeders move southwest in winter.
FOOD/HABITS Seeds of trees are the main food of siskins. Birds can be seen acrobatically searching for seeds in birch and alder catkins and in spruce cones. The siskin also feeds on nut-baskets in gardens. In winter, it moves around in flocks and is sometimes seen alongside redpolls.
BREEDING April to May. Cup-nests are built high in spruces. Three to five reddish-speckled pale blue eggs are incubated by the female for up to 14 days. Young fledge in about a fortnight. Two broods each season.

Redpoll
Carduelis flammea
SIZE AND DESCRIPTION 11.5–14 cm long. Its greyish-brown, dark-streaked colouring makes the redpoll look like a small sparrow, but the bill is small and broad. Both sexes have a red forehead and a small black bib. The adult male has a red upper breast. The wings are dark, with faint wing-bars. Juveniles lack the red forehead. Note the deeply forked tail.
VOICE The flight call is a hard metallic "chet-chet-chet". The song flight is the same metallic call interspersed with a buzzing sound.
HABITAT Breeds in young conifers, birch forest, willows on mountainsides and small dense copses in open country. Breeds mainly in Scandinavia and northern Russia, but also in the British Isles, Iceland and the mountains of central Europe. There is a southward movement in autumn, although redpolls do not winter further south than the Alps.
FOOD/HABITS Feeds in flocks, searching among the tips of birch and other trees. Flocks are often numerous and restless, moving from one site to another. Very nimble when feeding in branches. Visits gardens in winter.
BREEDING April to June. Nests in loose colonies, with nests of twigs and grasses built in trees and bushes. Four to five reddish-speckled pale blue eggs are incubated by the female for 10–13 days. Young fledge in 11–14 days. One or two broods a year.

Female

Male

Male

Female

House sparrow
Passer domesticus

SIZE AND DESCRIPTION 14–16 cm long. The male has a grey cap and grey breast, with an extensive black throat-patch. The brown back has dark streaks in both sexes. The female has a pale brown cap and buff eye-stripe. The wings of both sexes have small white wing-bars.

VOICE A number of monotonous chirps.

HABITAT The house sparrow seems to be completely linked to humans. It is found in towns, villages and farmland near human habitation. In winter, flocks can be seen feeding in fields. It does not occur in Iceland, except in Rekjavik, or in the mountains of northern Norway, but otherwise it is found across Europe, wherever there is a human presence.

FOOD/HABITS An omnivore that feeds on seeds and insects, as well as bread and other food left by people. Social, even when breeding.

BREEDING April to June. Builds an untidy nest in holes in buildings and sometimes in trees, including among the dense branches of Lawson's cypress. The three to five grey-blotched pale blue eggs are incubated for 11–14 days by the female. Young fledge within 15 days. Two or three broods per year.

Goldfinch
Carduelis carduelis

SIZE AND DESCRIPTION 12–13.5 cm long. With its red face, white cheeks and throat, black cap and black-and-gold wings, the goldfinch is a truly beautiful bird. In flight, the wings show their broad golden bands, and the white rump and black tail with its white markings are clearly visible. Sexes are alike, but the juvenile has a brown-streaked head until late summer or early autumn.

VOICE A piercing and cheerful trisyllabic "tickelitt" call. The song is rather soft, with a series of rapid trills and twitters involving the "tickelitt" call.

HABITAT Breeds in open lowland woodland, heaths, orchards and gardens, south from southern Scandinavia. East-European populations move south in winter.

FOOD/HABITS Feeds on seeds and berries, taking insects when feeding young. The pointed bill enables the goldfinch to extract seeds from thistleheads and teasels. Outside the breeding season, the goldfinch is seen in flocks.

BREEDING May to August. Nests are neat constructions of grass, moss and lichens, lined with thistledown or wool and built at the tips of branches. Four to seven dark-spotted blue eggs are incubated by the female for up to a fortnight. Young fledge in 13–16 days. Two broods each season.

Juvenile

113

Mammals

Although there are 4000 species of mammals in the world, there are only about 180 in Europe. Of these 30 are marine mammals and 23 are introductions from other parts of the world. The largest wild animals to visit the garden are mammals, but since most are nocturnal they are not often seen. Mammals are warm-blooded, air-breathing vertebrates. Their hearts, like those of birds, have four chambers. Body temperature is regulated internally and maintained by the body hair. Mammals usually bear live young, but there are half a dozen species of egg-laying mammals in Australia. All mammals suckle their young.

The most primitive mammals to be found in gardens are members of the order, Insectivora, which are small, insect-hunting ground-dwellers with long muzzles and continuous rows of teeth for crunching through insect chitin. Hedgehogs, moles and shrews are all insectivores.
Bats belong to the order, Chiroptera, and show some similarities with the insectivores, but the taxonomic link is not yet clear. Physically bats are unusual in that their forelimbs have become extended to carry the membranes of their wings. Identification of bats is best done in the hand, but it is possible to identify some species in the field, especially using bat detectors, which pick up their high frequency calls. Bats may roost in houses and it is illegal to disturb them in Britain.

Worldwide the largest order of mammals is Rodentia and several species are found in gardens. Rodents are characterised by the two pairs of front teeth, or incisors, which are long and grow continuously, becoming worn down by use. The lower incisors are particularly long and grow from extremely long sockets. Rodents lack canine teeth and there is a gap between the incisors and the grinding cheek-teeth. Squirrels are rodents. Red squirrels may be seen in wooded gardens in continental Europe and the grey squirrel, an introduction from North America, is a frequent garden visitor in Britain. Mice, voles and rats are all rodents and are predominantly vegetarian, although common rats will eat a wide variety of food including carrion. Another feature of rodents is their ability to breed rapidly and prolifically. They are prevented from becoming too numerous because of predation by birds of prey and by other mammals.

Predatory mammals belong to the order, Carnivora, which are characterised by their enlarged fangs, or canine teeth, and by the carnassial teeth on each jaw for cutting and shearing flesh. Foxes belong to the dog family and hunt small animals and search for carrion. The other carnivores likely to be found in gardens belong to the family, Mustelidae, and include weasels, stoats, martens and badgers. They all have prominent scent glands near the anus. An important carnivore in gardens is the domestic cat, which although not wild probably accounts for the deaths of more small mammals and birds than any of the wild predators.

Deer are not commonly seen in gardens, although in gardens near forests the roe deer and muntjac may be seen in England. Deer are even-toed ungulates, Artiodactyla, an order that covers most of the large herbivores, including cattle, pigs and hippotamuses.

1 Mole

Talpa europaea

SIZE AND DESCRIPTION Body is 11–16 cm long; tail measures 2–4 cm.
With its soft grey-black fur, cylindrical shape, massive earth-moving front
paws and pink, bewhiskered nose, the mole is very distinctive. Its tiny
eyes are covered by fur. The fur itself has a texture that allows it to lie
in any direction. The mole is rarely seen above ground, but the hills it
leaves are an indication of its presence.

HABITAT This woodland species has adapted to fields, parks and
gardens. It prefers soil that is well-drained and easy to dig. It is absent
from Iceland, Ireland, south-west Spain, Greece, much of Italy, Norway
and northern Sweden.

FOOD/HABITS Throughout the year, the mole is active beneath the
ground both day and night. Three- or four-hour sessions of activity
alternate with similar periods of rest and sleep. The mole's tunnels
may be 1 metre deep and 200 m long. Food is earthworms and insect
larvae, which it finds by smell and hearing. Cats, dogs, tawny owls,
stoats, green-keepers and keen gardeners are the mole's main enemies.

BREEDING The short breeding season begins from late February to early
June, depending on the latitude. Females are only in season for three
or four days, and males will mate with several females during a season.
Gestation lasts for about four weeks. The female builds a nest of dry
grass and leaves, in which she gives birth to three or four naked, blind
young. The young are suckled for four to five weeks, and leave the nest
at seven to nine weeks. Lifespan is about three years.

2 Hedgehog

Erinaceus europaeus

SIZE AND DESCRIPTION Body is 20–30 cm long; tail measures 1–4 cm. The
rounded, rather short body is covered with spines, which are dark with
creamy tips. The face and undersides are covered with coarse hairs. The
hedgehog has longish legs, and each foot has five long toes. Its nose is
pointed, the ears and eyes are small, and the teeth are pointed.

HABITAT Found from lowlands up to 2,000 m, where there is ground
cover for shelter and nesting. Occurs across Europe, from southern
Scandinavia and Finland. Has been introduced to a number of islands.

FOOD/HABITS Usually nocturnal, the hedgehog finds prey by sound and
scent. It eats almost any invertebrates found at ground level, including
slugs, worms and beetles, as well as bird eggs and nestlings and carrion.
It runs quite swiftly, can climb banks and walls, and can also swim. Mainly
solitary, with males ranging up to 3 km each night. Spines protect the
hedgehog from most predators, but foxes and badgers may eat them.
Cars, lawn-mowers and poisoning by chemicals, such as slug pellets, are
the main causes of death. Hibernation begins in October, with younger
animals still active in November, and ends in March or April.

BREEDING Males seek females shortly after emerging from hibernation.
Each partner disperses and may mate with other individuals. Gestation
takes about five weeks. Two to seven young are born blind and suckled
for about four weeks. Females may produce a second litter. Males
take no part in rearing the family. Mortality is high in the first year, but
survivors will live for about three years.

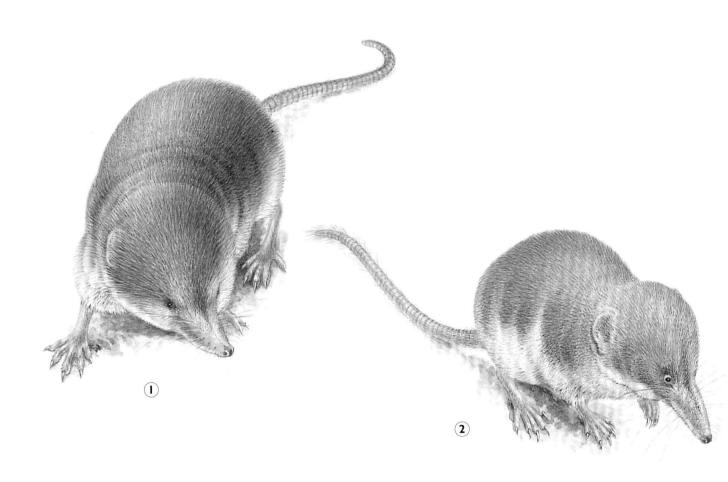

① ②

1 Common shrew

Sorex araneus

SIZE AND DESCRIPTION Body is 5.2–8.7 cm long; tail measures 2.4–4.4 cm. The adult has a three-coloured coat: dark brown on the head and back, flanks of pale brown and grey-brown underparts. Juveniles have a light-brown back and a hairy tail. The long, whiskered nose is much longer than a mouse's.

HABITAT Widespread in habitats with ground cover, but most common in rough grassland, scrub, woodland and hedges. Found across Europe from Scandinavia to Wales, but replaced on the continent west of Belgium with the very similar species, Millet's shrew. Absent from Iceland, Ireland, Spain and Portugal.

FOOD/HABITS Active both day and night. Using its long nose to sniff out beetles, spiders, small snails and other invertebrates, the common shrew is only active for about 30 minutes at a time. It returns to its nest and sleeps for a few minutes, before feeding on cached food and grooming itself. It then begins to hunt again. To survive, it must eat about 90% of its weight each day. Fiercely territorial outside the breeding season.

BREEDING The breeding season lasts from March to September. Gestation is 13–19 days, and there may be five litters per season. Four to ten young are born, naked and blind. Weaning takes 22 days. Less than 30% survive long enough to breed the following year and few of these survive until the next breeding season. A lactating female will need to eat one-and-a-half times its own weight each day.

2 Pygmy shrew

Sorex minutus

SIZE AND DESCRIPTION Body is 4–6.4 cm long; tail measures 3.2–4.6 cm. Distinguished from the juvenile common shrew by its two-coloured coat: back and head are grey-brown and underside is greyish-white. The head looks more bulbous and the tail is thicker and hairier than the common shrew's. The teeth are reddish (a useful characteristic to note on dead specimens).

HABITAT Similar to the common shrew. Can be found in gardens, particularly near compost heaps.

FOOD/HABITS Very agile, and often climbs in search of food. Does not burrow itself, but will use burrows of other small mammals and may dig through leaf litter and surface vegetation. It feeds on insects and other invertebrates of 2–6 mm in length. Earthworms are usually too large, but small slugs and snails will be eaten. It eats one-and-a-quarter times its own weight each day. Tawny and barn owls are the pygmy shrew's main predators, but domestic cats and stoats are also enemies.

BREEDING The breeding season lasts from April to October, with peak breeding in mid-summer. After a gestation of 22–25 days, four to seven young are born, naked and blind. They suckle for about 22 days. Several litters in a season. Pygmy shrews live for up to 13 months, but at least half do not survive their first two months.

1 Wood mouse

Apodemus sylvaticus

SIZE AND DESCRIPTION Body is 8–10 cm long; tail measures 6.9–11.5 cm. The wood mouse has orange-brown fur, noticeable ears and a tail that may be longer than its body. Underparts are pale grey.

HABITAT Found in every habitat, except those that are too wet or too high (above 2,500 m). Rare in coniferous woodland, where suitable food is scarce. Distributed across Europe including Iceland, but excluding much of Norway and Sweden and all of Finland.

FOOD/HABITS Forages largely at night for seeds, buds, fruits, nuts, snails and spiders. A good climber, it can be found in trees and will enter houses in search of food. It does not hibernate, but slows down in cold weather. Male home ranges are 0.25–0.33 ha; female ranges are half the size.

BREEDING Males compete for females where their territories overlap. Mating is from March to September. Gestation lasts 19–26 days and results in litters of 2–11 young, which are blind for up to 16 days and weaned after 22 days. Females may have six litters in a season. Males are mature at seven weeks, females at six weeks. Life expectancy is 10–17 weeks, but a few reach two years.

2 House mouse

Mus domesticus

SIZE AND DESCRIPTION Body is 7–10 cm long; tail measures 6.5–10 cm. Uniformly greyish fur, with a thick, scaly tail. Colour varies, and mice seen on London Underground tracks are darker. It is from the house mouse that pet mice have been bred. A strong musky smell is emitted when they are disturbed.

HABITAT Having originated in rocky habitats in Asia, this mouse is now closely associated with humans. It is found in buildings and on farmland throughout Europe, except for mountains in the far north of Norway.

FOOD/HABITS Primarily a grain-eater, the house mouse feeds on a wide variety of seeds, roots, fungi, and invertebrates, taking an average of about 3.5 grammes per day.

BREEDING Gestation is 19–20 days and litters have five to eight young, whose eyes open within three days. Young are weaned by 18 days, and independent in about four weeks. Sexual maturity is reached in about 45 days. Five to ten litters per year. Mortality is high, and very few survive to reach two years.

3 Yellow-necked mouse

Apodemus flavicollis

SIZE AND DESCRIPTION Body is 8.5–13 cm long; tail measures 9–13.5 cm. Larger, brighter and redder than the wood mouse, the yellow-necked mouse has a clearer demarcation between its upperside and underside. There is a distinct yellow bib between the front legs in northern Europe, but this is less distinct in the south of its range.

HABITAT Favourite habitats are deciduous woodland, wooded gardens and established orchards. It is also found in coniferous forests and mountains up to 2,250 m. Found in Wales, southern England, northern Spain, France, but absent from northern Scandinavia.

FOOD/HABITS Eats acorns, seeds, grain, fungi, berries, leaves, insect larvae and pupae, snails and birds' eggs. Will hoard food throughout the year. A good climber. Largely nocturnal, and will travel further when foraging than the wood mouse. Territories vary from 0.2–5 ha.

BREEDING Breeding starts in February. Litters of 2–11 young are born after a gestation of 12–16 days. The young are blind for up to 16 days, but become independent after 21 days. Females are sexually mature at about eight weeks. Three or four litters per year. Lifespan of four years is possible.

1 Bank vole
Clethrionomys glareolus
SIZE AND DESCRIPTION Body is 8–11 cm long; tail measures 3.5–7.2 cm. The bank vole has a rounded face and small eyes that are less noticeable than in other voles. Bright chestnut fur on the back, with buffish grey underparts.
HABITAT Found in mixed deciduous woodland with well-developed undergrowth, but also in coniferous woodland, grassland and scrub. Occurs across Europe from northern Spain to northern Scandinavia, but is absent from much of Ireland, Iceland, Italy and Greece. Very common, but enormous population variations in continental Europe.
FOOD/HABITS Actively searches for food during the day in summer, with peaks of activity around dawn and dusk. Seeds, roots, nuts, shoots and buds are all eaten.
BREEDING The breeding season starts in April and may continue into December. Gestation is 18–20 days, and average litter size is three to five young. Eyes open after about 12 days, and the young are independent at 21 days. Sexual maturity is reached in about six weeks. Four or five litters each year. Lifespan rarely exceeds 18 months.

2 Field vole
Microtus agrestis
SIZE AND DESCRIPTION Body is 9.5–13.5 cm long; tail measures 2.5–4.6 cm. The field vole has coarse, yellowish grey-brown fur and a short tail. Its ears are hidden by fur. Similar to the common vole, which is found on the European mainland.
HABITAT Prefers rough grassland and scrub. Also found in open woodland, field margins and hedges. Across Europe from northern Spain, but absent from Iceland, Ireland, Italy, the Balkans, Norway and Sweden.
FOOD/HABITS Mostly nocturnal, but sometimes active during the day, particularly in winter. Almost entirely vegetarian, eating leaves and seeds.
BREEDING The breeding season lasts from March to October, and may continue longer in mild winters. Gestation is 18–20 days, and litters contain two to seven young. The eyes of the young voles open at nine days. Young are independent at three weeks. Females are sexually mature at six weeks. A female may have up to seven litters each year. Only a few reach 18 months.

3 Brown rat
Rattus norvegicus
SIZE AND DESCRIPTION Body is 11–29 cm long; tail measures 8.5–23 cm. The brown rat has coarse grey-brown fur and a thick, scaly tail. Occasionally melanistic individuals occur, but otherwise there is no other colour variation. The black or ship rat is similar, but it is darker and has a longer, more slender tail.
HABITAT Found across Europe, except in Arctic Scandinavia, in all habitats associated with man. Will move into farmland in summer in search of cereals.
FOOD/HABITS This wary rodent is nocturnal and is rarely seen by human beings, despite living in close proximity to them. Prefers grain, but will eat a wide variety of food including roots, buds, fruit, frogs, birds' eggs and human refuse.
BREEDING It breeds throughout the year. A third of females are pregnant at any one time. Gestation is 21–24 days. Seven or eight young are born. They are blind for about a week, but independent by six weeks and reach sexual maturity at 11–12 weeks. Most females produce about five litters a year. Lifespan one year.

1 Edible dormouse
Glis glis

SIZE AND DESCRIPTION Body is 13–19 cm long; tail measures 11–15 cm. Uniformly grey, with a faint yellowish tinge, a fleshy nose and naked, rounded ears. The eyes are surrounded by dark rings, making them seem larger. The tail is less bushy than grey squirrel's.

HABITAT Mature woodlands, parks and large gardens. Across Europe, but absent from Scandinavia. Introduced to Britain.

FOOD/HABITS Nocturnal and secretive, it spends the day hidden in a hole in a tree or in an old bird's nest. It forages in the tree canopy, rarely on the ground, but will enter sheds and lofts in search of food. Nuts, berries are its favourite foods, but some insects and nestlings are also eaten. Hibernation begins in October, with up to eight individuals hibernating together until April or May.

BREEDING Mating occurs from June to August. Gestation is 28 days. A litter consists of four to six young. Lifespan up to six years.

2 Red squirrel
Sciurus vulgaris

SIZE AND DESCRIPTION Body is 1–25 cm long; tail measures 14–20 cm. Two colour phases exist on the continent: dark greyish-brown and russet red. Only the red phase is found in the British Isles. In winter, the fur becomes greyer and the ear-tufts are prominent. In the British subspecies, the tail can be very pale in summer. Smaller than the grey squirrel.

HABITAT Forests, especially coniferous, but also in woods dominated by beeches. Found up to 2,000 m in the Alps and Pyrenees. Distributed across Europe from northern Spain to northern Russia.

FOOD/HABITS A solitary, diurnal animal, with peaks of activity around dawn and dusk. Feeds high in the canopy on seeds in cones, berries and fruit. Will eat insects, nestling birds and eggs. Hoards food.

BREEDING Mating varies between winter and spring, according to availability of food and geographical distribution. Gestation is 36–42 days. Three to eight young per litter. Eyes open at 28–30 days, and young become independent at about 11 weeks. Sexually mature at 10–12 months. Lifespan three to five years but many die in first year.

3 Grey squirrel
Sciurus carolinensis

SIZE AND DESCRIPTION Body is 23–30 cm long; tail measures 19–24 cm. The grey fur is variably tinged with red and yellow. Some individuals can be noticeably red in summer. The rounded ears are never tufted. Apart from the red squirrel, the only other confusable species is the smaller, more mouse-like edible dormouse.

HABITAT The grey squirrel is a north American species that has been introduced to Britain and Ireland, where it is now very common in wooded habitats, including gardens and parks.

FOOD/HABITS Active during the day. In summer, foraging takes place mainly in trees, but it will also search on the ground for fungi, bulbs, roots and cached acorns. Food includes eggs, nestlings, leaves, buds and shoots. The drey is a large structure of twigs, leaves, bark and grass. It does not hibernate, and can only survive three days without feeding.

BREEDING Mating usually takes place in May and December. Males approach a female making noisy, chattering sounds. Up to seven young are born, naked and blind. Lifespan can be up to nine years.

1 Weasel
Mustela nivalis
SIZE AND DESCRIPTION Body is 13–23 cm long; tail measures 3–6 cm. Males are much larger than females. The long body and neck give the weasel a snake-like appearance. Fur is chestnut brown, with white underparts. It lacks the black-tipped tail of the larger stoat.
HABITAT Found in lowland woods, farmland and large gardens across Europe, but absent from Ireland and Iceland.
FOOD/HABITS Hunts day and night, with three periods of sleep (three to four hours each) every day. Rests in burrows, rabbit warrens and badger setts. Prey is mainly voles, but includes young rabbits, rats, moles, and nestlings. Cannot survive more than 24 hours without food. Solitary outside breeding season.
BREEDING Mating occurs February to August. Gestation is 34–37 days. Three to eight young are born. They are weaned at six to eight weeks, and independent at 12 weeks. Females are sexually mature at three or four months, and may breed in their first year, if food is plentiful. Life expectancy is about a year in the wild.

2 Stoat
Mustela erminea
SIZE AND DESCRIPTION Body is 24–31 cm with 9–14 cm tail. Larger than weasel with more reddish-brown fur and black-tipped tail. In the north of its range it becomes white except for the tip of its tail. Males are larger than females.
HABITAT Has a wide range of habitats and is found wherever there is suitable food. Found throughout Europe except for the lowlands around the Mediterranean and Iceland.
FOOD/HABITS Moves with a bounding gait and is a good swimmer and agile climber. Hunts by both day and night either singly or in family parties. Prey includes voles, mice, rabbits, birds and hares, which are usually killed by a sharp bite at the back of the neck.
BREEDING Although mating takes place in summer, the gestation, of 21–28 days, is delayed until the following spring, when a litter of six to 12 is born. Eyes of the young open at 5–6 weeks, black tail-tip appears at 6 or 7 weeks. Weaning at 5 weeks. Independent at about 10 weeks. Life expectancy in the wild is between one and two years.

3 Pine marten
Martes martes
SIZE AND DESCRIPTION Body is 40–55 cm long; tail measures 22–26 cm. A cat-sized mammal with a flattened head, long neck and short legs. Fur is predominantly dark brown, but the throat is cream or pale yellow. The beech marten is similar, but has a white throat.
HABITAT Woodland, particularly coniferous forests, up to 2,000 m. Across Europe from northern Spain, but in British Isles it is limited to the north of Scotland, the Lake District, north Wales, Yorkshire and western Ireland. It moves into lofts and farm buildings in winter.
FOOD/HABITS Nocturnal. A good climber and jumper. Hunts on the ground by scent, taking voles, squirrels, rabbits, hares and mice, birds' eggs and nestlings. It also eats berries, fruit and the honey of wild bees.
BREEDING Mating occurs in July or August. Implantation in January. Gestation is about 28 days. Usually three young are born in March or April. Young are weaned at 45 days and remain with the family until they disperse in August. If the first winter is survived, life expectancy is probably about five years.

1 Red fox
Vulpes vulpes

SIZE AND DESCRIPTION Body is 56–77 cm long; tail measures 28–49 cm;
height at shoulder is 35–40 cm. The pointed nose, pointed ears and
bushy tail make this reddish-brown animal, which is the size of a large
domestic cat, unmistakable. Males weigh about 15% more than females.
Some individuals may be quite pale, and others rather dark.

HABITAT If there is sufficient cover, red foxes will be found in every type
of habitat. Up to 3,500 m in mountains. Widespread across Europe,
they are only absent from Iceland. They are now found in parks, gardens
and industrial areas of cities.

FOOD/HABITS Active mostly at night, with peak activity at dawn and
dusk, the fox preys on rabbits, hares, rats, voles and ground-nesting
birds, including domestic hens and ducks. About two-thirds of an urban
fox's diet is human refuse. It also feeds on carrion. Earthworms are an
important food for cubs. Hedgerow berries and fruit are also eaten.
Foxes live in family groups, made up of a dog fox, a dominant vixen and
up to six other females, usually offspring from the previous year. The
size of the family territory varies according to the availability of food.

BREEDING The trisyllabic call of foxes in mid-winter is a sign of the
mating season. When the female and male make contact they stay
together for a few days, during which time the dog guards the vixen
from the attention of other males. Gestation is about 52 days. Four
or five cubs are born, blind and black. They are fully weaned at seven
weeks, but begin to catch insects and earthworms at about five weeks.
Most cubs disperse in late August. Although some may survive for 12
years, most foxes live for about two years.

2 Badger
Meles meles

SIZE AND DESCRIPTION Body is 67–85 cm long; tail measures 11–20 cm;
height at shoulder 30 cm. With the black stripe through each eye, its
white face and sheer bulk, the badger is unlike any other mammal likely
to be found in a European garden. The coarse fur is greyish. Although
sexes are alike, the male has a broad, domed head, while the female's is
narrower and flatter. Females have bushier tails.

HABITAT The favoured habitat is deciduous woodland with open
areas or bordering farmland, but it is also found in parks, gardens and
mountains up to 2,000 m. Occurs across Europe, but is absent from
Iceland, Sardinia, Corsica and northern Scandinavia.

FOOD/HABITS Badgers eat whatever is available. Earthworms are
important, and up to 200 may be consumed in a single night. Insects,
small mammals, birds, hedgehogs, eggs, snakes, lizards and frogs are all
eaten, as are berries, fruit, roots and the honey and grubs in wasps' and
bees' nests.

BREEDING Mating occurs throughout the year, but is mostly from
February to May. Because of delayed implantation, most cubs are
born in January to March, with a peak in February. A litter consists of
one to five cubs, which are weaned at about 12 weeks. They become
independent at 24 weeks. If they survive their first 12 months, badgers
in the wild will live for about six years.

Mammals

1 Muntjac
Muntiacus reevesi

SIZE AND DESCRIPTION Body is 90–107 cm long; tail measures 14–18 cm; height at shoulder is 44–52 cm. This small red-brown deer is about the size of a Labrador retriever. Males have short antlers, which create distinct ridges along their brows. Both sexes have dark V-shapes on their heads. Bucks have long incisors, which can occasionally be seen protruding from the upper lip. Muntjacs are short-legged, and the way they hold their heads makes them look rather hump-backed. In autumn, they take on a grey-brown coat and the legs become almost black. The underside of the tail is white and only noticeable when the tail is lifted. Roe deer are taller and have a black nose. The muntjac's droppings are cylindrical, blackish-brown and about 1 cm long.
HABITAT The muntjac was introduced to England from China in the 1900s, and is now widespread throughout the south of England. Dense deciduous woodland with thick undergrowth is the preferred habitat, but it is also found in gardens and orchards.
FOOD/HABITS The normally solitary muntjac is active both at day and night, particularly around dusk and dawn. Grass is eaten regularly during the spring, but the muntjac is usually a browser on shrubs and the lower branches of deciduous trees. Herbs are eaten in spring. The muntjac is a serious problem in ancient woodlands. It uses regular pathways.
BREEDING Females come into season a few days after giving birth. One fawn is born after a gestation of seven months. It is weaned after 17 weeks. Muntjacs are frequent road casualties, fawns may be taken by foxes, some are killed by dogs and others are shot. Many will die in hard winters, but survivors may live for over 16 years.

2 Roe deer
Capreolus capreolus

SIZE AND DESCRIPTION Body is 100–140 cm long; tail measures 1–2 cm; height at shoulder is 100–140 cm. The roe deer is taller and longer-legged than the muntjac. Its coat is reddish-brown in summer, moulting into a longer grey-brown (sometimes almost black) winter coat in September or October. The short muzzle has a noticeable black nose. The eyes are large and dark. Bucks have small antlers, seldom longer than 30 cm and with three tines. Droppings are dark brown and round.
HABITAT Found in deciduous and coniferous woodland, open moorland and sometimes in reedbeds. Will also enter large gardens. Distributed across Europe from south-west England and Scotland, and from Norway and Sweden to south of the Iberian peninsula.
FOOD/HABITS Feeds on buds and shoots of trees and shrubs, brambles, wild flowers, ivy, ferns and berries. Mostly solitary. May form small flocks in winter, and males may be seen with females in the breeding season.
BREEDING Bucks are territorial from April to August. Rutting takes place in July and August, but implantation is delayed and gestation does not start until December or January. Young are born from April to June. Most have twins, which are weaned after two to three months. Most young die in their first few months, but the natural lifespan of survivors is 12–14 years.

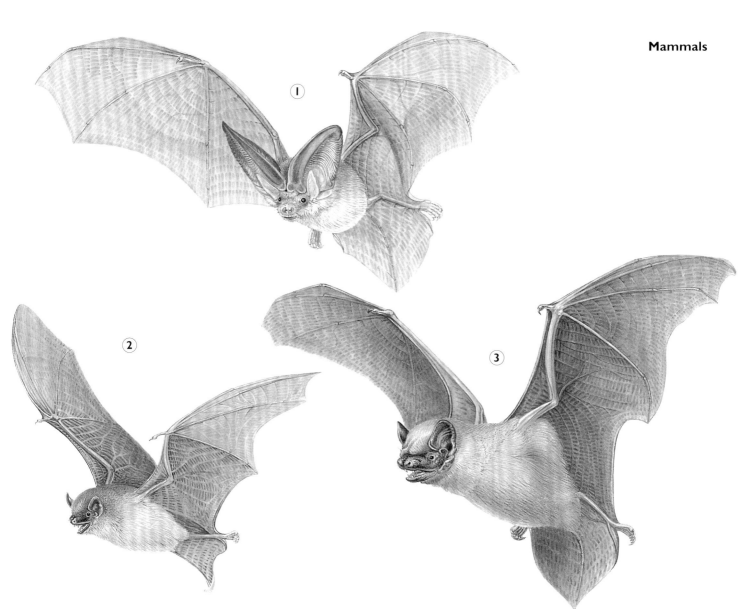

1 Brown long-eared bat
Plecotus auritus
SIZE AND DESCRIPTION Body is 3–4.3 cm long; wingspan is 23–28.5 cm. Not a large bat, but with prominent ears that are clearly visible in flight. Fluffy fur is greyish-brown, with a paler underside. The brown wings are translucent.
HABITAT Found in mature woodlands, parkland and large gardens up to 2,000 m. Absent from Spain, Iceland and northern Scandinavia.
FOOD/HABITS Summer roosts, with up to 60 individuals, are found in old trees, buildings and batboxes. Winter roosts are in caves, tunnels and disused mines, and the bats may move to a new roost in trees in mid-winter. They are active in the roost for about 90 minutes before emerging about 30 minutes after dusk. The bats hibernate from late October to early April.
BREEDING Mating takes place from October to April. One or two young are born in mid-June and July, and weaned at six weeks. Average lifespan is four and a half years.

2 Pipistrelle
Pipistrellus pipistrellus
SIZE AND DESCRIPTION Body is 3.5–4.9 cm long; wingspan is 27–30 cm. A very small bat, the pipistrelle has a soft reddish coat, but colours may vary. The rounded head has small triangular ears. In flight, the wings are narrow. Fast and jerky flight.
HABITAT Found across Europe, in all but the most exposed habitats. Absent from Iceland and all but the southernmost parts of Norway, Sweden and Denmark.
FOOD/HABITS In summer, pipistrelles roost in buildings, squeezing through tiny gaps to gain entrance. In winter, they use both buildings and natural sites for hibernation. They will roost behind boards attached to walls and in specially made boxes. Roosts are indicated by a pile of cylindrical droppings about 0.5 cm long. They usually emerge about 20 minutes after sunset, but may be seen during daylight. Ranging up to 2 km from the roost, they hunt flying insects for up to three hours in warm weather, and for 15 minutes maximum in cold.
BREEDING Mating takes place from August to November. Each male has a group of eight to ten females. Implantation is delayed and lasts for 44 days, with usually one young being born in spring. Average lifespan is four years.

3 Noctule
Nyctalus noctula
SIZE AND DESCRIPTION Body is 3–8 cm long; wingpan is 32–40 cm. Almost twice the size of the pipistrelle, with narrow wings. Coat is golden-brown, but moults into a duller, paler brown in August or September. The wings are dark brown or black.
HABITAT Lowland, deciduous woodland, parkland, and gardens with mature trees. Absent from Iceland, Scotland and much of Scandinavia.
FOOD/HABITS Tree-holes are used as summer roosts, as are batboxes. In winter, both trees and buildings are roosted. Noctules emerge from roosts at dusk. Flying insects are caught and eaten on the wing.
BREEDING Mating is in September and October, but may also take place during winter. Gestation lasts for 70–73 days, with offspring born in June or July. Usually only one young is produced. It is independent at seven weeks. It may live for 12 years but most die much sooner.

Index

Index